Non-parametric Tests for Complete Data

# Non-parametric Tests for Complete Data

Vilijandas Bagdonavičius
Julius Kruopis
Mikhail S. Nikulin

First published 2011 in Great Britain and the United States by ISTE Ltd and John Wiley & Sons, Inc.

ISTE Ltd
27-37 St George's Road
London SW19 4EU
UK

www.iste.co.uk

John Wiley & Sons, Inc.
111 River Street
Hoboken, NJ 07030
USA

www.wiley.com

© ISTE Ltd 2011

The rights of Vilijandas Bagdonaviçius, Julius Kruopis and Mikhail S. Nikulin to be identified as the authors of this work have been asserted by them in accordance with the Copyright, Designs and Patents Act 1988.

Library of Congress Cataloging-in-Publication Data

Bagdonavicius, V. (Vilijandas)
  Nonparametric tests for complete data / Vilijandas Bagdonavicius, Julius Kruopis, Mikhail Nikulin.
    p. cm.
  Includes bibliographical references and index.
  ISBN 978-1-84821-269-5 (hardback)
  1. Nonparametric statistics. 2. Statistical hypothesis testing. I. Kruopis, Julius. II. Nikulin, Mikhail (Mikhail S.) III. Title.
  QA278.8.B34 2010
  519.5--dc22

                                                                          2010038271

British Library Cataloguing-in-Publication Data
A CIP record for this book is available from the British Library
ISBN 978-1-84821-269-5

Printed and bound in Great Britain by CPI Antony Rowe, Chippenham and Eastbourne.

# Table of Contents

# Preface

Testing hypotheses in non-parametric models are discussed in this book. A statistical model is non-parametric if it cannot be written in terms of a finite-dimensional parameter. The main hypotheses tested in such models are hypotheses on the probability distribution of elements of the following: data homogeneity, randomness and independence hypotheses. Tests for such hypotheses from complete samples are considered in many books on non-parametric statistics, including recent monographs by Maritz [MAR 95], Hollander and Wolfe [HOL 99], Sprent and Smeeton [SPR 01], Govindarajulu [GOV 07], Gibbons and Chakraborti [GIB 09] and Corder and Foreman [COR 09].

This book contains tests from complete samples. Tests for censored samples can be found in our book *Tests for Censored Samples* [BAG 11].

In Chapter 1, the basic ideas of hypothesis testing and general hypotheses on non-parametric models are briefly described.

In the initial phase of the solution of any statistical problem the analyst must choose a model for data analysis. The correctness of the data analysis strongly depends on the choice of an appropriate model. Goodness-of-fit tests are used to check the adequacy of a model for real data.

One of the most-applied goodness-of-fit tests are chi-squared type tests, which use grouped data. In many books on statistical data analysis, chi-squared tests are applied incorrectly. Classical chi-squared tests are based on theoretical results which are obtained assuming that the ends of grouping intervals do not depend on the sample, and the parameters are estimated using grouped data. In real applications, these assumptions are often forgotten. The modified chi-squared tests considered in Chapter 2 do not suffer from such drawbacks. They are based on the assumption that the ends of grouping intervals depend on the data, and the parameters are estimated using initially non-grouped data.

Another class of goodness-of-fit tests based on functionals of the difference of empirical and theoretical cumulative distribution functions is described in Chapter 3. The tests for composite hypotheses classical statistics are modified by replacing unknown parameters by their estimators. Application of these tests is often incorrect because the critical values of the classical tests are used in testing the composite hypothesis and applying modified statistics.

In section 5.5, special goodness-of-fit tests which are not from the two above-mentioned classes, and which are specially designed for specified probability distributions, are given.

Tests for the equality of probability distributions (homogeneity tests) of two or more independent or dependent random variables are considered in several chapters. Chi-squared type tests are given in section 2.5 and tests based on functionals of the difference of empirical distribution functions are given in section 3.5. For many alternatives, the most efficient tests are the rank tests for homogeneity given in sections 4.4 and 4.6–4.8.

Classical tests for the independence of random variables are given in sections 2.4 (tests of chi-square type), and 4.3 and 5.2 (rank tests).

Tests for data randomness are given in sections 4.3 and 5.2.

All tests are described in the following way: 1) a hypothesis is formulated; 2) the idea of test construction is given; 3) a statistic on which a test is based is given; 4) a finite sample and (or) asymptotic distribution of the test statistic is found; 5) a test, and often its modifications (continuity correction, data with *ex aequo*, various approximations of asymptotic law) are given; 6) practical examples of application of the tests are given; and 7) at the end of the chapters problems with answers are given.

Anyone who uses non-parametric methods of mathematical statistics, or wants to know the ideas behind and mathematical substantiation of the tests, can use this book. It can be used as a textbook for a one-semester course on non-parametric hypotheses testing.

Knowledge of probability and parametric statistics are needed to follow the mathematical developments. The basic facts on probability and parametric statistics used in the the book are also given in the appendices.

The book consists of five chapters, and appendices. In each chapter, the numbering of theorems, formulas, and comments are given using the chapter number.

The book was written using lecture notes for graduate students in Vilnius and Bordeaux universities.

We thank our colleagues and students at Vilnius and Bordeaux universities for comments on the content of this book, especially Rūta Levulienė for writing the computer programs needed for application of the tests and solutions of all the exercises.

Vilijandas BAGDONAVIČIUS
Julius KRUOPIS
Mikhail NIKULIN

# Terms and Notation

$||\boldsymbol{A}||$ – the norm $(\sum_i \sum_j a_{ij}^2)^{1/2}$ of a matrix $\boldsymbol{A} = [a_{ij}]$;

$\boldsymbol{A} > \boldsymbol{B}$ $(\boldsymbol{A} \geq \boldsymbol{B})$ – the matrix $\boldsymbol{A} - \boldsymbol{B}$ is positive (non-negative) definite;

$a \vee b$ $(a \wedge b)$ – the maximum (the minimum) of the numbers $a$ and $b$;

ASE – the asymptotic relative efficiency;

$B(n,\ p)$ – binomial distribution with parameters $n$ and $p$;

$B^-(n,\ p)$ – negative binomial distribution with parameters $n$ and $p$;

$Be(\gamma,\ \eta)$ – beta distribution with parameters $\gamma$ and $\eta$;

cdf – the cumulative distribution function;

CLT – the central limit theorem;

$\mathbf{Cov}(X,\ Y)$ – the covariance of random variables $X$ and $Y$;

$\mathbf{Cov}(\mathbf{X},\ \boldsymbol{Y})$ – the covariance matrix of random vectors $\mathbf{X}$ and $\boldsymbol{Y}$;

$\mathbf{E}X$ – the mean of a random variable $X$;

$\mathbf{E}(\mathbf{X})$ – the mean of a random vector $\mathbf{X}$;

$\mathbf{E}_\theta(X)$, $\mathbf{E}(X|\theta)$, $\mathbf{Var}_\theta(X)$, $\mathbf{Var}(X|\theta)$ – the mean or the variance of a random variable $X$ depending on the parameter $\theta$;

$\mathcal{E}(\lambda)$ – exponential distribution with parameters $\lambda$;

$F(m,\ n)$ – Fisher distribution with $m$ and $n$ degrees of freedom;

$F(m,\ n;\ \delta)$ – non-central Fisher distribution with $m$ and $n$ degrees of freedom and non-centrality parameter $\delta$;

$F_\alpha(m,\ n)$ – $\alpha$ critical value of Fisher distribution with $m$ and $n$ degrees of freedom;

$F_T(x)$ $(f_T(x))$ – the cdf (the pdf) of the random variable $T$;

$f(x;\ \theta)$, $f(x|\theta)$ – the pdf depending on a parameter $\theta$;

$F(x;\ \theta)$, $F(x|\theta)$ – the cdf depending on a parameter $\theta$;

$G(\lambda,\ \eta)$ – gamma distribution with parameters $\lambda$ and $\eta$;

iid – independent identically distributed;

$LN(\mu,\ \sigma)$ – lognormal distribution with parameters $\mu$ and $\sigma$;

LS – least-squares (method, estimator);

ML – maximum likelihood (function, method, estimator);

$N(0,\ 1)$ – standard normal distribution;

$N(\mu,\ \sigma^2)$ – normal distribution with parameters $\mu$ and $\sigma^2$;

$N_k(\boldsymbol{\mu}, \ \boldsymbol{\Sigma})$ – $k$-dimensional normal distribution with the mean vector $\boldsymbol{\mu}$ and the covariance matrix $\boldsymbol{\Sigma}$;

$\mathcal{P}(\lambda)$ – Poisson distribution with a parameter $\lambda$;

pdf – the probability density function;

$\mathbf{P}\{A\}$ – the probability of an event $A$;

$\mathbf{P}\{A|B\}$ – the conditional probability of event $A$;

$\mathbf{P}_\theta\{A\}$, $\mathbf{P}\{A|\theta\}$ – the probability depending on a parameter $\theta$;

$\mathcal{P}_k(n, \ \boldsymbol{\pi})$ – $k$-dimensional multinomial distribution with parameters $n$ and $\boldsymbol{\pi} = (\pi_1, ..., \pi_k)^T$, $\pi_1 + ... + \pi_k = 1$;

rv – random variable

$S(n)$ – Student's distribution with $n$ degrees of freedom;

$S(n; \ \delta)$ – non-central Student's distribution with $n$ degrees of freedom and non-centrality parameter $\delta$;

$t_\alpha(n)$ – $\alpha$ critical value of Student's distribution with $n$ degrees of freedom;

$U(\alpha, \ \beta)$ – uniform distribution in the interval $(\alpha, \ \beta)$;

UMP – uniformly most powerful (test);

UUMP – unbiased uniformly most powerful (test);

$\mathbf{Var}\,X$ – the variance of a random variable $X$;

$\mathbf{Var}(\mathbf{X})$ – the covariance matrix of a random vector $\mathbf{X}$;

$W(\theta, \ \nu)$ – Weibull distribution with parameters $\theta$ ir $\nu$;

$X, \ Y, \ Z, ...$ – random variables;

$\mathbf{X}$, $\mathbf{Y}$, $\mathbf{Z}$, ... – random vectors;

$\mathbf{X}^T$ – the transposed vector, i.e. a vector-line;

$||\boldsymbol{x}||$ – the length $(\boldsymbol{x}^T\boldsymbol{x})^{1/2} = (\sum_i x_i^2)^{1/2}$ of a vector $\boldsymbol{x} = (x_1, ..., x_k)^T$;

$X \sim N(\mu, \sigma^2)$ – random variable $X$ normally distributed with parameters $\mu$ and $\sigma^2$ (analogously in the case of other distributions);

$X_n \overset{P}{\to} X$ – convergence in probability $(n \to \infty)$;

$X_n \overset{a.s.}{\to} X$ – almost sure convergence or convergence with probability 1 $(n \to \infty)$;

$X_n \overset{d}{\to} X$, $F_n(x) \overset{d}{\to} F(x)$ – weak convergence or convergence in distribution $(n \to \infty)$;

$X_n \overset{d}{\to} X \sim N(\mu, \sigma^2)$ – random variables $X_n$ asymptotically $(n \to \infty)$ normally distributed with parameters $\mu$ and $\sigma^2$;

$X_n \sim Y_n$ – random variables $X_n$ and $Y_n$ asymptotically $(n \to \infty)$ equivalent $(X_n - Y_n \overset{P}{\to} 0)$;

$x(P)$ – $P$-th quantile;

$x_P$ – $P$-th critical value;

$z_\alpha$ – $\alpha$ critical value of the standard normal distribution;

$\Sigma = [\sigma_{ij}]_{k \times k}$ – covariance matrix;

$\chi^2(n)$ – chi-squared distribution with $n$ degrees of freedom;

$\chi^2(n; \delta)$ – non-central chi-squared distribution with $n$ degrees of freedom and non-centrality parameter $\delta$;

$\chi_\alpha^2(n)$ – $\alpha$ critical value of chi-squared distribution with $n$ degrees of freedom.

# Chapter 1

# Introduction

## 1.1. Statistical hypotheses

The simplest model of statistical data is a *simple sample*, i.e. a vector $\mathbf{X} = (X_1, ..., X_n)^T$ of $n$ independent identically distributed random variables. In real experiments the values $x_i$ of the random variables $X_i$ are observed (measured). The non-random vector $x = (x_1, ..., x_n)^T$ is a realization of the simple sample $\mathbf{X}$.

In more complicated experiments the elements $X_i$ are dependent, or not identically distributed, or are themselves random vectors. The random vector $\mathbf{X}$ is then called a *sample*, not a simple sample.

Suppose that the cumulative distribution function (cdf) $F$ of a sample $\mathbf{X}$ (or of any element $X_i$ of a simple sample) belongs to a set $\mathcal{F}$ of cumulative distribution functions. For example, if the sample is simple then $\mathcal{F}$ may be the set of absolutely continuous, discrete, symmetric, normal, Poisson cumulative distribution functions. The set $\mathcal{F}$ defines a *statistical model*.

Suppose that $\mathcal{F}_0$ is a subset of $\mathcal{F}$.

The statistical hypothesis $H_0$ is the following assertion: *the cumulative distribution function $F$ belongs to the set $\mathcal{F}_0$.* We write $H_0 : F \in \mathcal{F}_0$.

The hypothesis $H_1 : F \in \mathcal{F}_1$, where $\mathcal{F}_1 = \mathcal{F} \backslash \mathcal{F}_0$ is the complement of $\mathcal{F}_0$ to $\mathcal{F}$ is called *alternative* to the hypothesis $H_0$.

If $\mathcal{F} = \{F_{\boldsymbol{\theta}}, \boldsymbol{\theta} \in \Theta \subset \mathbf{R}^m\}$ is defined by a finite-dimensional parameter $\boldsymbol{\theta}$ then the model is *parametric*. In this case the statistical hypothesis is a statement on the values of the finite-dimensional parameter $\boldsymbol{\theta}$.

In this book non-parametric models are considered. A statistical model $\mathcal{F}$ is called *non-parametric* if $\mathcal{F}$ is *not defined by a finite-dimensional parameter*.

If the set $\mathcal{F}_0$ contains only one element of the set $\mathcal{F}$ then the hypothesis is *simple*, otherwise the hypothesis is *composite*.

## 1.2. Examples of hypotheses in non-parametric models

Let us look briefly and informally at examples of the hypotheses which will be considered in the book. We do not formulate concrete alternatives, only suppose that models are non-parametric. Concrete alternatives will be formulated in the chapters on specified hypotheses.

### 1.2.1. Hypotheses on the probability distribution of data elements

The first class of hypotheses considered in this book consists of hypotheses on the form of the cdf $F$ of the elements of a sample.

Such hypotheses may be simple or composite.

A simple hypothesis has the form $H_0 : F = F_0$; here $F_0$ is a specified cdf. For example, such a hypothesis may mean that the $n$ numbers generated by a computer are realizations of random variables having uniform $U(0, 1)$, Poisson $\mathcal{P}(2)$, normal $N(0, 1)$ or other distributions.

A composite hypothesis has the form $H_0 : F \in \mathcal{F}_0 = \{F_{\boldsymbol{\theta}}, \boldsymbol{\theta} \in \Theta\}$, where $F_{\boldsymbol{\theta}}$ are cdfs of known analytical form depending on the finite-dimensional parameter $\boldsymbol{\theta} \in \Theta$. For example, this may mean that the salary of the doctors in a city are normally distributed, or the failure times of TV sets produced by a factory have the Weibull distribution.

More general composite hypotheses, meaning that the data verify some parametric or semi-parametric regression model, may be considered. For example, in investigating the influence of some factor $z$ on the survival time the following hypothesis on the cdf $F_i$ of the $i$-th sample element may be used:

$$F_i(x) = 1 - \{1 - F_0(x)\}^{\exp\{\beta z_i\}}, i = 1, \ldots, n$$

where $F_0$ is an unknown baseline cdf, $\beta$ is an unknown scalar parameter and $z_i$ is a known value of the factor for the $i$-th sample element.

The following tests for simple hypotheses are considered: *chi-squared tests* (section 2.2) and *tests based on the difference of empirical and cumulative distribution functions* (sections 3.2 and 3.3).

The following tests for composite hypotheses are considered: general tests such as *chi-squared tests* (sections 2.3 and 2.4, *tests based on the difference of non-parametric and parametric estimators of the cumulative distribution function* (section 3.4), and also *special tests* for specified families of probability distributions (section 5.5).

### 1.2.2. *Independence hypotheses*

Suppose that $(X_i, Y_i)^T$, $i = 1, 2...., n$ is a simple sample of the random vector $(X, Y)^T$ with the cdf $F = F(x, y) \in \mathcal{F}$; here $\mathcal{F}$ is a non-parametric class two-dimensional cdf.

An *independence hypothesis* means that the components $X$ and $Y$ are independent. For example, this hypothesis may mean that the sum of sales of managers $X$ and the number of complaints from consumers $Y$ are independent random variables.

The following tests for independence of random variables are considered: *chi-squared independence tests* (section 2.5) and *rank tests* (sections 4.3 and 4.10).

### 1.2.3. *Randomness hypothesis*

A randomness hypothesis means that the observed vector $x = (x_1, ..., x_n)^T$ is a realization of a simple sample $\mathbf{X} = (X_1, ..., X_n)^T$, i.e. of a random vector with independent and identically distributed (iid) components.

The following tests for randomness hypotheses are considered: *runs tests* (section 5.2) and *rank tests* (section 4.4).

### 1.2.4. *Homogeneity hypotheses*

A homogeneity hypothesis of two independent simple samples $\mathbf{X} = (X_1, ..., X_m)^T$ and $\mathbf{Y} = (Y_1, ..., Y_n)^T$ means that the cdfs $F_1$ and $F_2$ of the random variables $X_i$ and $Y_j$ coincide. The homogeneity hypothesis of $k > 2$ independent samples is formulated analogously.

The following tests for homogeneity of independent simple samples are considered: *chi-squared tests* (section 2.6), *tests based on the difference of cumulative distribution functions* (section 3.5), *rank tests* (sections 4.5 and 4.8), and some *special tests* (section 5.1).

If $n$ independent random vectors $\mathbf{X}_i = (X_{i1}, ..., X_{ik})^T$, $i = 1, ..., n$ are observed then the vectors $(X_{1j}, ..., X_{nj})^T$ composed of the components are $k$ dependent samples, $j = 1, ..., k$. The homogeneity hypotheses of $k$ related samples means the equality of the cdfs $F_1, ..., F_k$ of the components $X_{i1}, ..., X_{ik}$.

The following tests for homogeneity of related samples are considered: *rank tests* (sections 4.7 and 4.9) and other special tests (sections 5.1, 5.3 and 5.4).

### 1.2.5. *Median value hypotheses*

Suppose that $\mathbf{X} = (X_1, ..., X_n)^T$ is a simple sample of a continuous random variable $X$. Denote by $M$ the median of the random variable $X$. The median value hypothesis has the form $H : M = M_0$; here $M_0$ is a specified value of the median.

The following tests for this hypothesis are considered: *sign tests* (section 5.1) and *rank tests* (section 4.6).

## 1.3. Statistical tests

A *statistical test* or simply a *test* is a rule which enables a decision to be made on whether or not the zero hypothesis $H_0$ should be rejected on the basis of the observed realization of the sample.

Any test considered in this book is based on the values of some statistic $T = T(\mathbf{X}) = T(X_1, ..., X_n)$, called the *test statistic*. Usually the statistic $T$ takes different values under the hypothesis $H_0$ and the alternative $H_1$. If the statistic $T$ has a tendency to take smaller (greater) values under the hypothesis $H_0$ than under the alternative $H_1$ then the hypothesis $H_0$ is *rejected* in favor of the alternative if $T > c$ ($T < c$, respectively), where $c$ is a well-chosen real number.

If the values of the statistic $T$ have a tendency to concentrate in some interval under the hypothesis and outside this interval under the alternative then the hypothesis $H_0$ is *rejected* in favor of the alternative if $T < c_1$ or $T > c_2$, where $c_1$ and $c_2$ are well-chosen real numbers.

Suppose that the hypothesis $H_0$ is rejected if $T > c$ (the other two cases are considered similarly).

The probability

$$\beta(F) = \mathbf{P}_F\{T > c\}$$

of rejecting the hypothesis $H_0$ when the true cumulative distribution function is a specified function $F \in \mathcal{F}$ is called the *power function* of the test. When using a test, two types of error are possible:

1. The hypothesis $H_0$ is rejected when it is true, i.e. when $F \in \mathcal{F}_0$. Such an error is called a *type I error*. The probability of this error is $\beta(F)$, $F \in \mathcal{F}_0$.

2. The hypothesis $H_0$ is not rejected when it is false, i.e. when $F \in \mathcal{F}_1$. Such an error is called a *type II error*. The probability of this error is $1 - \beta(F)$, $F \in \mathcal{F}_1$.

The number

$$sup_{F \in \mathcal{F}_0} \beta(F) \tag{1.1}$$

is called the *significance level* of the test .

Fix $\alpha \in (0, 1)$. If the significance level does not exceed $\alpha$ then for any $F \in \mathcal{F}_0$ the type I error does not exceed $\alpha$.

Usually tests with significance level values not greater than $\alpha = 0.1; 0.05; 0.01$ are used.

If the distribution of the statistic $T$ is absolutely continuous then, usually, for any $\alpha \in (0, 1)$ we can find a test based on this statistic such that the significance level is equal to $\alpha$.

A test with a significance level not greater than $\alpha$ is called *unbiased*, if

$$inf_{F \in \mathcal{F}_1} \beta(F) \geq \alpha \qquad [1.2]$$

This means that the zero hypothesis is rejected with greater probability under any specified alternative than under the zero hypothesis. Let $\mathcal{T}$ be a class of test statistics of unbiased tests with a significance level not greater than $\alpha$.

The statistic $T$ defines the *uniformly most powerful unbiased* test in the class $\mathcal{T}$ if $\beta_T(F) \geq \beta_{T^*}(F)$ for all $T^* \in \mathcal{T}$ and for all $F \subset \mathcal{F}_1$.

A test is called *consistent* if for all $F \in \mathcal{F}_1$

$$\beta(F) \rightarrow 1, \quad \text{as } n \rightarrow \infty \qquad [1.3]$$

This means that if $n$ is large then under any specified alternative the probability of rejecting the zero hypothesis is near to 1.

## 1.4. *P*-value

Suppose that a simple statistical hypothesis $H_0$ is rejected using tests of one of the following forms:

1) $T \geq c$; 2) $T \leq c$; or 3) $T \leq c_1$ or $T \geq c_2$; here $T = T(\mathbf{X})$ is a test statistic based on the sample $\mathbf{X} = (X_1, \ldots, X_n)^T$.

We write $\mathbf{P}_0\{A\} = \mathbf{P}\{A|H_0\}$.

Fix $\alpha \in (0, 1)$. The first (second) test has a significance level not greater than $\alpha$, and nearest to $\alpha$ if the constant $c = \inf\{s : \mathbf{P}_0\{T \geq s\} \leq \alpha\}$ $(c = \sup\{s : \mathbf{P}_0\{T \leq s\} \leq \alpha\})$.

The third test has a significance level not greater than $\alpha$ if $c_1 = \sup\{s : \mathbf{P}\{T \leq s\} \leq \alpha/2\}$ and $c_2 = \inf\{s : \mathbf{P}\{T \geq s\} \leq \alpha/2\}$.

Denote by $t$ the observed value of the statistic $T$. In the case of the tests of the first two forms the $P$-values are defined as the probabilities

$$pv = \mathbf{P}_0\{T \geq t\} \quad \text{and} \quad pv = \mathbf{P}_0\{T \leq t\}.$$

Thus the $P$-value is the probability that under the zero hypothesis $H_0$ the statistic $T$ takes a value more distant than $t$ in the direction of the alternative (in the first case to the right, in the second case to the left, from $t$).

In the third case, if

$$\mathbf{P}_0\{T \leq t\} \leq \mathbf{P}_0\{T \geq t\}$$

then

$$pv/2 = \mathbf{P}_0\{T \leq t\}$$

and if

$$\mathbf{P}_0\{T \leq t\} \geq \mathbf{P}_0\{T \geq t\}$$

then

$$pv/2 = \mathbf{P}_0\{T \geq t\}$$

So in the third case the $P$-value is defined as follows

$$pv = 2\min\{\mathbf{P}_0\{T \leq t\}, \mathbf{P}_0\{T \geq t\}\} = 2\min\{F_T(t), 1 - F_T(t-)\}$$

where $F_T$ is the cdf of the statistic $T$ under the zero hypothesis $H_0$. If the distribution of $T$ is absolutely continuous and symmetric with respect to the origin, the last formula implies

$$pv = 2\min\{F_T(t), F_T(-t)\} = 2F_T(-|t|) = 2\{1 - F_T(|t|)\}$$

If the result observed during the experiment is a rare event when the zero hypothesis is true then the $P$-value is small and the hypothesis should be rejected. This is confirmed by the following theorem.

**Theorem 1.1.** *Suppose that the test is of any of the three forms considered above. For the experiment with the value $t$ of the statistic $T$ the inequality $pv \leq \alpha$ is equivalent to the rejection of the zero hypothesis.*

**Proof.** Let us consider an experiment where $T = t$. If the test is defined by the inequality $T \geq c$ ($T \leq c$) then $c = \inf\{s : \mathbf{P}_0\{T \geq s\} \leq \alpha\}$ ($c = \sup\{s : \mathbf{P}_0\{T \leq s\} \leq \alpha\}$) and $\mathbf{P}_0\{T \leq t\} = pv$ ($\mathbf{P}_0\{T \geq t\} = pv$). So the inequality $pv \leq \alpha$ is equivalent to the inequality $t \geq c$ ($t \leq c$). The last inequalities mean that the hypothesis is rejected.

If the test is defined by the inequalities $T \leq c_1$ or $T \geq c_2$ then $c_1 = \sup\{s : \mathbf{P}_0\{T \leq s\} \leq \alpha/2\}$, $c_2 = \inf\{s : \mathbf{P}_0\{T \geq s\} \leq \alpha/2\}$ and $2\min\{\mathbf{P}_0\{T \leq t\}, \mathbf{P}_0\{T \geq t\}\} = pv$. So the inequality $pv \leq \alpha$ means that $2\min\{\mathbf{P}_0\{T \leq t, \mathbf{P}_0\{T \geq t\}\} \leq \alpha$.

If $\mathbf{P}_0\{T \leq t\} \geq \mathbf{P}_0\{T \geq t\}$, then the inequality $pv \leq \alpha$ means that $\mathbf{P}_0\{T \geq t\} \leq \alpha/2$. This is equivalent to the inequality $t \geq c_2$, which means that the hypothesis is rejected. Analogously, if $\mathbf{P}_0\{T \leq t\} \geq \mathbf{P}_0\{T \geq t\}$ then the inequality $pv \leq \alpha$ means that $\mathbf{P}_0\{T \leq t\} \leq \alpha/2$. This is equivalent to the inequality $t \leq c_1$, which means that the hypothesis is rejected. So in both cases the inequality $pv \leq \alpha$ means that the hypothesis is rejected.

$\triangle$

If the critical region is defined by the asymptotic distribution of $T$ (usually normal or chi-squared) then the $P$-value $pv_a$ is computed using the asymptotic distribution of $T$, and it is called the *asymptotic $P$-value.*

Sometimes the $P$-value $pv$ is interpreted as random because each value $t$ of $T$ defines a specific value of $pv$. In the case of the alternatives considered above the $P$-values are the realizations of the following random variables:

$$1 - F_T(T-), \quad F_T(T) \quad \text{and} \quad 2\min\{F_T(T), 1 - F_T(T-)\}$$

## 1.5. Continuity correction

If the distribution of the statistic $T$ is discrete and the asymptotic distribution of this statistic is absolutely continuous (usually normal) then for medium-sized samples the approximation of distribution $T$ can be improved using the *continuity correction* [YAT 34].

The idea of a continuity correction is explained by the following example.

**Example 1.1.** Let us consider the parametric hypothesis: $H$ : $p = 0.5$ and the alternative $H_1 : p > 0.5$; here $p$ is the Bernoulli distribution parameter. For example, suppose that during $n = 20$ Bernoulli trials the number of successes is $T = 13$. It is evident that the hypothesis $H_0$ is rejected if the statistic $T$ takes large values, i.e. if $T \geq c$ for a given $c$. Under $H_0$, the statistic $T$ has the binomial distribution $B(20; 0.5)$. The exact $P$-value is

$$pv = \mathbf{P}\{T \geq 13\} = \sum_{i=13}^{20} C_{20}^i (1/2)^{20} = I_{1/2}(13, \, 8) = 0.131588$$

Using the normal approximation

$$Z_n = (T - 0.5n)/\sqrt{0.25n} = (T - 10)/\sqrt{5} \overset{d}{\to} Z \sim N(0, \, 1)$$

we obtain the asymptotic $P$-value

$$pv_a = \mathbf{P}\{T \geq 13\} = \mathbf{P}\{\frac{T - 10}{\sqrt{5}} \geq \frac{13 - 10}{\sqrt{5}}\} \approx$$

$$1 - \Phi(\frac{13 - 10}{\sqrt{5}}) = 0.089856$$

This is considerably smaller than the exact $P$-value.

Note that $\mathbf{P}\{T \geq 13\} = \mathbf{P}\{T > 12\}$. If we use the normal approximation then the same probability may be approximated by:

$$1 - \Phi((13 - 10)/\sqrt{5}) = 0.089856$$

$$\text{or by} \quad 1 - \Phi((12 - 10)/\sqrt{5}) = 0.185547$$

Both approximations are far from the exact $P$-value. The continuity correction is therefore performed using the normal approximation in the center 12.5 of the interval $(12, 13]$. So the *asymptotic P-value with a continuity correction* is

$$pv_{cc} = 1 - \Phi((13 - 0.5 - 10)/\sqrt{5}) = 0.131776$$

The obtained value is very similar to the exact $P$-value.

In the case of the alternative $H_2 : p < 0.5$ the zero hypothesis is rejected if $T \leq d$, and the $P$-value is

$$pv = \mathbf{P}\{T \leq 13\} = \sum_{i=0}^{13} C_{20}^i (1/2)^2 0 = I_{1/2}(7, \ 14) = 0.942341$$

In this case

$$pv_a = \Phi((13 - 10)/\sqrt{5}) = 0.910144$$

Note that $\mathbf{P}\{T \leq 13\} = \mathbf{P}\{T < 14\}$. If we use the normal approximation, the same probability is approximated by

$$\Phi((13 - 10)/\sqrt{5}) = 0.910144 \quad \text{or} \quad \Phi((14 - 10)/\sqrt{5}) = 0.963181$$

Both are far from the exact $P$-value. So the continuity correction is performed using the normal approximation in the middle 13.5 of the interval $(13, 14]$. Therefore the asymptotic $P$-value with a continuity correction is

$$pv_{cc} = \Phi((13 + 0.5 - 10)/\sqrt{5}) = 0.941238$$

The obtained value is very similar to the exact $P$-value.

In the case of the bilateral alternative $H_3 : p \neq 0.5$ the exact and asymptotic $P$-values are

$$pv = 2\min\{F_T(13), 1 - F_T(13-)\}$$

$$= 2\min(0.942341; 0.131588) = 0.263176$$

and $pv_a = 2\min(0.910144; 0.089856) = 0.179712$, respectively. The asymptotic $P$-value with a continuity correction is

$$pv_{cc} = 2\min(\Phi((13 + 0.5 - 10)/\sqrt{5}), 1 - \Phi((13 - 0.5 - 10)/\sqrt{5})) =$$

$$2\min(0.941238, 0.131776) = 0.263452$$

Generalizing, suppose that the test statistic $T$ takes integer values and under the zero hypothesis the asymptotic distribution of the statistic

$$Z = \frac{T - \mathbf{E}T}{\sqrt{\mathbf{Var}T}}$$

is standard normal. If the critical region is defined by the inequalities

$$a)\ T \geq c; \quad b)\ T \leq c; \quad c)\ T \leq c_1 \quad \text{or} \quad T \geq c_2;$$

and the observed value of the statistic $T$ is $t$ then the asymptotic $P$-values with a continuity correction are

$$pv_{cc} = 1 - \Phi((t - 0.5 - \mathbf{E}T)/\sqrt{\mathbf{Var}T})$$

$$pv_{cc} = \Phi((t + 0.5 - \mathbf{E}T)/\sqrt{\mathbf{Var}T})$$

$$pv_{cc} = 2\min\left[\Phi\left(\frac{t + 0.5 - \mathbf{E}T}{\sqrt{\mathbf{Var}T}}\right), 1 - \Phi\left(\frac{t - 0.5 - \mathbf{E}T}{\sqrt{\mathbf{Var}T}}\right)\right] \quad [1.4]$$

respectively.

## 1.6. Asymptotic relative efficiency

Suppose that under the zero hypothesis and under the alternative the distribution of the data belongs to a non-parametric family depending on a scalar parameter $\theta$ and possibly another parameter $\vartheta$. Let us consider the hypothesis $H_0 : \theta = \theta_0$ with the one-sided alternatives $H_1 : \theta > \theta_0$ or $H_2 : \theta < \theta_0$ and the two-sided alternative $H_3 : \theta \neq \theta_0$.

**Example 1.2.** Let $\mathbf{X} = (X_1, ..., X_n)^T$ and $\mathbf{Y} = (Y_1, ..., Y_m)^T$ be two independent simple samples, $X_i \sim F(x)$ and $Y_j \sim F(x - \theta)$, where $F(x)$ is an unknown absolutely continuous cdf (the parameter $\vartheta$) and $\theta$ is a location parameter. Under the homogeneity hypothesis $\theta = 0$, under the one-sided alternatives $\theta > 0$ or $\theta < 0$, and under the two-sided alternative $\theta \neq 0$. So the homogeneity hypothesis can be formulated in terms of the scalar parameter: $H_0 : \theta = 0$. The possible alternatives are $H_1 : \theta > 0$, $H_2 : \theta < 0$, $H_3 : \theta \neq 0$.

Let us consider the one-sided alternative $H_1$. Fix $\alpha \in (0, 1)$. Suppose that the hypothesis is rejected if

$$T_n > c_{n,\alpha}$$

where $n$ is the sample size and $T_n$ is the test statistic.

Denote by

$$\beta_n(\theta) = \mathbf{P}_\theta(T_n > c_{n,\alpha}\}$$

the power function of the test.

Most of the tests are consistent, so the power of such tests is close to unity if the sample size is large. So the limit of the power under fixed alternatives is not suitable for comparing the performance of different tests.

To compare the tests the behavior of the powers of these tests under the sequence of approaching alternatives

$$H_n : \theta = \theta_n = \theta_0 + \frac{h}{n^\delta}, \quad \delta > 0, \ h > 0,$$

may be compared.

Suppose that the same sequence of approaching alternatives is written in two ways

$$\theta_m = \theta_0 + \frac{h_1}{n_{1m}^\delta} = \theta_0 + \frac{h_2}{n_{2m}^\delta}$$

where $n_{im} \to \infty$ as $m \to \infty$, and

$$\lim_{m \to \infty} \beta_{n_{1m}}(\theta_m) = \lim_{m \to \infty} \beta_{n_{2m}}(\theta_m)$$

Then the limit (if it exists and does not depend on the choice of $\theta_m$)

$$e(T_{1n}, T_{2n}) = \lim_{m \to \infty} \frac{n_{2m}}{n_{1m}}$$

is called the *asymptotic relative efficiency* (ARE) [PIT 48] of the first test with respect to the second test.

The ARE is the inverse ratio of sample sizes necessary to obtain the same power for two tests with the same asymptotic significance level, while simultaneously the sample sizes approach infinity and the sequence of alternatives approaches $\theta_0$.

Under regularity conditions, the ARE have a simple expression.

**Regularity assumptions:**

1) $\mathbf{P}_{\theta_0}\{T_{in} \geq c_{n,\alpha}\} \to \alpha.$

2) In the neighborhood of $\theta_0$ there exist

$$\mu_{in}(\theta) = \mathbf{E}_\theta T_{in}, \quad \sigma_{in}(\theta) = \mathbf{Var}_\theta T_{in}$$

and the function $\mu_{in}(\theta)$ is infinitely differentiable at the point $\theta_0$; furthermore, $\dot\mu_{in}(\theta_0) > 0$, and higher order derivatives are equal to 0; $i = 1, 2$.

3) There exists

$$\lim_{n\to\infty} \mu_{in}(\theta) = \mu_i(\theta), \quad \lim_{n\to\infty} n^\delta \sigma_{in}(\theta) = \sigma_i(\theta), \quad \mu_i(\theta_0)/\sigma_i(\theta_0) > 0$$

where $\delta > 0$.

4) For any $h \geq 0$

$$\dot\mu_{in}(\theta_n) \to \dot\mu_i(\theta_0), \quad \sigma_{in}(\theta_n) \to \sigma_i(\theta_0), \quad \text{as} \quad n \to \infty$$

5) Test statistics are asymptotically normal:

$$\mathbf{P}_{\theta_n}\{(T_{in} - \mu_{in}(\theta_n))/\sigma_{in}(\theta_n) \leq z\} \to \Phi(z).$$

**Theorem 1.2.** *If the regularity assumptions are satisfied then the asymptotic relative efficiency can be written in the form*

$$e(T_{1n}, T_{2n}) = \left( \frac{\dot\mu_1(\theta_0)/\sigma_1(\theta_0)}{\dot\mu_2(\theta_0)/\sigma_2(\theta_0)} \right)^{1/\delta} \qquad [1.5]$$

**Proof.** First let us consider one statistic and skip the indices $i$. Let us find $\lim_{n\to\infty} \beta_n(\theta_n)$. By assumption 1

$$\mathbf{P}_{\theta_0}\{T_n > c_{n,\alpha}\} = \mathbf{P}_{\theta_0}\{\frac{T_n - \mu_n(\theta_0)}{\sigma_n(\theta_0)} > \frac{c_{n,\alpha} - \mu_n(\theta_0)}{\sigma_n(\theta_0)}\} \to \alpha$$

so

$$z_{n,\alpha} = \frac{c_{n,\alpha} - \mu_n(\theta_0)}{\sigma_n(\theta_0)} \to z_\alpha$$

By assumptions 2–4

$$\frac{\mu_n(\theta_n) - \mu_n(\theta_0)}{\sigma_n(\theta_0)} = \frac{\dot{\mu}_n(\theta_0)hn^{-\delta} + o(1)}{n^{-\delta}\sigma(\theta_0) + o(1)} \rightarrow \frac{\dot{\mu}(\theta_0)}{\sigma(\theta_0)}h$$

So using assumption 5 we have

$$\beta_n(\theta_n) = \mathbf{P}_{\theta_n}\{T_n > c_{n,\alpha}\} = \mathbf{P}_{\theta_n}\{\frac{T_n - \mu_n(\theta_n)}{\sigma_n(\theta_n)} > \frac{c_{n,\alpha} - \mu_n(\theta_n)}{\sigma_n(\theta_n)}\} =$$

$$\mathbf{P}_{\theta_n}\{\frac{T_n - \mu_n(\theta_n)}{\sigma_n(\theta_n)} > z_{n,\alpha}\frac{\sigma_n(\theta_0)}{\sigma_n(\theta_n)} - \frac{\mu_n(\theta_n) - \mu_n(\theta_0)}{\sigma_n(\theta_0)}\frac{\sigma_n(\theta_0)}{\sigma_n(\theta_n)}\}$$

$$\rightarrow 1 - \Phi\left(z_\alpha - h\frac{\dot{\mu}(\theta_0)}{\sigma(\theta_0)}\right)$$

Let $T_{1n}$ and $T_{2n}$ be two test statistics verifying the assumptions of the theorem and let

$$\theta_m = \theta_0 + \frac{h_1}{n_{1m}^\delta} = \theta_0 + \frac{h_2}{n_{2m}^\delta}$$

be a sequence of approaching alternatives. The last equalities imply

$$\frac{h_2}{h_1} = (\frac{n_{2m}}{n_{1m}})^\delta$$

We have proved that

$$\beta_{n_{im}}(\theta_m) \rightarrow 1 - \Phi\left(z_\alpha - h_i\frac{\dot{\mu}_i(\theta_0)}{\sigma_i(\theta_0)}\right)$$

Choose $n_{1m}$ and $n_{2m}$ to give the same limit powers. Then

$$\frac{n_{2m}}{n_{1m}} = (\frac{h_2}{h_1})^{1/\delta} = \left(\frac{\dot{\mu}_1(\theta_0)/\sigma_1(\theta_0)}{\dot{\mu}_2(\theta_0)/\sigma_2(\theta_0)}\right)^{1/\delta}$$

$\triangle$

# Chapter 2

# Chi-squared Tests

## 2.1. Introduction

Chi-squared tests are used when data are classified into several groups and only numbers of objects belonging to concrete groups are used for test construction. The vector of such random numbers has a multinomial distribution and depends only on the finite number of parameters. So chi-squared tests, being based on this vector, are parametric but are also used for non-parametric hypotheses testing, so we include them in this book.

When the initial data are replaced by grouped data, some information is lost, so this method is used when more powerful tests using all the data are not available.

## 2.2. Pearson's goodness-of-fit test: simple hypothesis

Suppose that $\mathbf{X} = (X_1, ..., X_n)^T$ is a simple sample of a random variable $X$ having the c.d.f. $F$ from a non-parametric class $\mathcal{F}$.

**Simple hypothesis**

$$H_0 : F(x) = F_0(x), \quad \forall x \in \mathbf{R} \tag{2.1}$$

where $F_0$ is completely specified (known) cdf from the family $\mathcal{F}$.

The hypotheses

$$H_0 : X \sim U(0, 1), \quad H_0 : X \sim B(1, 0.5), \quad H_0 : X \sim N(0, 1)$$

are examples of simple non-parametric hypotheses. For example, such a hypothesis is verified if we want to know whether realizations generated by a computer are obtained from the uniform $U(0, 1)$, Poisson $\mathcal{P}(2)$, normal $N(0, 1)$ or other completely specified distribution.

The data are grouped in the following way: the abscissas axis is divided into a finite number of intervals using the points $-\infty = a_0 < a_1 < ... < a_k = \infty$. Denote by $U_j$ the number of $X_i$ falling in the interval $(a_{j-1}, \ a_j]$

$$U_j = \sum_{i=1}^{n} \mathbf{1}_{(a_{j-1}, a_j]}(X_i), \quad j = 1, 2...., k$$

So, instead of the fully informative data **X**, we use the *grouped data*

$$\boldsymbol{U} = (U_1, \ldots, U_k)^T$$

We can also say that the statistic $U$ is obtained using a special data *censoring mechanism*, known as the *mechanism of grouping data*. The random vector $U$ has the multinomial distribution $\mathcal{P}_k(n, \ \boldsymbol{\pi})$: for $0 \leq m_i \leq n$, $\sum_i m_i = n$

$$\mathbf{P}\{U_1 = m_1, ..., U_k = m_k\} = \frac{n!}{m_1!...m_k!} \pi_1^{m_1}...\pi_k^{m_k} \tag{2.2}$$

where $\pi_i = \mathbf{P}\{X \in (a_{i-1}, a_i]\} = F(a_i) - F(a_{i-1})$ is the probability that the random variable $X$ takes a value in the interval $(a_{i-1}, a_i]$, $\boldsymbol{\pi} = (\pi_1, \ldots, \pi_k)^T$, $\pi_1 + ... + \pi_k = 1$.

Under the hypothesis $H_0$, the following hypothesis also holds.

**Hypothesis on the values of multinomial distribution parameters**

$$H_0' : \pi_j = \pi_{j0}, \quad j = 1, 2...., k \qquad [2.3]$$

where $\pi_{j0} = F_0(a_j) - F_0(a_{j-1})$.

Under the hypothesis $H_0'$

$$U \sim \mathcal{P}_k(n, \ \boldsymbol{\pi}_0)$$

where $\boldsymbol{\pi}_0 = (\pi_{10}, \ldots, \pi_{k0})^T$, $\quad \pi_{10} + \ldots + \pi_{k0} = 1$.

If the hypothesis $H_0'$ is rejected then it is natural also to reject the narrower hypothesis $H_0$.

Pearson's chi-squared test for the hypothesis $H_0'$ is based on the differences between the maximum likelihood estimators $\hat{\pi}_j$ of the probabilities $\pi_j$ obtained from the grouped data $U$ and the hypothetical values $\pi_{j0}$ of these probabilities.

The relation $\sum_{j=1}^{k} \pi_j = 1$ implies that the multinomial model $\mathcal{P}_k(1, \ \boldsymbol{\pi})$ depends on $(k-1)$-dimensional parameters $(\pi_1, ..., \pi_{k-1})^T$.

From [2.2] the likelihood function of the random vector $U$ is

$$L(\pi_1, ..., \pi_{k-1}) = \frac{n!}{U_1!...U_k!} \pi_1^{U_1}...\pi_k^{U_k} \qquad [2.4]$$

The loglikelihood function is

$$\ell(\pi_1, ..., \pi_{k-1}) = \sum_{j=1}^{k-1} U_j \ln \pi_j + U_k \ln(1 - \sum_{j=1}^{k-1} \pi_j + C)$$

so

$$\ell_j = \frac{U_j}{\pi_j} - \frac{U_k}{1 - \sum_{j=1}^{k-1} \pi_j} = \frac{U_j}{\pi_j} - \frac{U_k}{\pi_k}$$

which implies that for all $j, l = 1, \ldots k$

$$U_j \pi_l = U_l \pi_j$$

Summing both sides with respect to $l$ and using the relations $\sum_{j=1}^{k} \pi_j = 1$ and $\sum_{j=1}^{k} U_j = n$, we obtain $U_j = n\pi_j$, so the maximum likelihood estimators of the parameters $\pi_i$ are

$$\hat{\boldsymbol{\pi}} = (\hat{\pi}_1, \ldots, \hat{\pi}_k)^T, \quad \hat{\pi}_i = U_i/n, \quad i = 1, \ldots, k$$

The famous *Pearson's statistic* has the form

$$X_n^2 = \sum_{i=1}^{k} \frac{(\sqrt{n}(\hat{\pi}_i - \pi_{i0}))^2}{\pi_{i0}} = \sum_{i=1}^{k} \frac{(U_i - n\pi_{i0})^2}{n\pi_{i0}} = \frac{1}{n} \sum_{i=1}^{k} \frac{U_i^2}{\pi_{i0}} - n.$$

$$[2.5]$$

Under the hypothesis $H_0'$, the realizations of the differences $\hat{\pi}_i - \pi_{i0}$ are scattered around zero. If the hypothesis is not true then at least one number $i$ exists such that the realizations of the differences $\hat{\pi}_i - \pi_{i0}$ are scattered around some positive or negative value. In such a case the statistic $X_n^2$ has a tendency to take greater values than under the zero hypothesis.

Pearson's test is asymptotic, i.e. it uses an asymptotic distribution of the test statistic $X_n^2$, which is chi square, which follows from the following theorem.

**Theorem 2.1.** *If* $0 < \pi_{i0} < 1$, $\pi_{10} + \cdots + \pi_{k0} = 1$ *then under the hypothesis* $H_0'$

$$X_n^2 \xrightarrow{d} \chi_{k-1}^2 \quad as \quad n \to \infty$$

**Proof.** Under the hypothesis $H_0'$, the random vector $\boldsymbol{U} = (U_1, \ldots, U_k)^T$ is the sum of iid random vectors $\boldsymbol{X}_i$ having the mean $\pi_0$ and the covariance matrix $\boldsymbol{D} = [d_{jj'}]_{k\times k}$; $d_{jj} = \pi_{j0}(1 - \pi_{j0})$, $d_{jj'} = -\pi_{j0}\pi_{j'0}$, $j \neq j'$.

If $0 < \pi_{i0} < 1$, $\pi_{10} + \cdots + \pi_{k0} = 1$ then the central limit theorem holds

$$\sqrt{n}(\hat{\boldsymbol{\pi}} - \boldsymbol{\pi}_0) = \sqrt{n}(\hat{\pi}_1 - \pi_{10}, \ldots, \hat{\pi}_k - \pi_{k0})^T \xrightarrow{d} \boldsymbol{Y} \sim N_k(\boldsymbol{0}, \ \boldsymbol{D})$$
[2.6]

as $n \to \infty$. The matrix $\mathbf{D}$ can be written in the form

$$\mathbf{D} = \boldsymbol{p}_0 - \boldsymbol{p}_0 \boldsymbol{p}_0^T$$

where $\boldsymbol{p}_0$ is the diagonal matrix with elements $\pi_{10}, \ldots, \pi_{k0}$ on the main diagonal. Set

$$\boldsymbol{Z}_n = \sqrt{n} \boldsymbol{p}_0^{-1/2}(\hat{\boldsymbol{\pi}} - \boldsymbol{\pi}_0) = \left( \frac{\sqrt{n}(\hat{\pi}_1 - \pi_{10})}{\sqrt{\pi_{10}}}, \ldots, \frac{\sqrt{n}(\hat{\pi}_k - \pi_{k0})}{\sqrt{\pi_{k0}}} \right)^T$$

The result [2.6] implies that

$$\boldsymbol{Z}_n \xrightarrow{d} \boldsymbol{Z} \sim N_k(\boldsymbol{0}, \ \boldsymbol{\Sigma})$$

where

$$\boldsymbol{\Sigma} = \boldsymbol{p}_0^{-1/2} \mathbf{D} \boldsymbol{p}_0^{-1/2} = \mathbf{E}_k - \mathbf{q}\mathbf{q}^T$$

where $\mathbf{q} = (\sqrt{\pi_{10}}, \ldots, \sqrt{\pi_{k0}})^T$, $\mathbf{q}^T \mathbf{q} = 1$, and $\mathbf{E}_k$ is a $k \times k$ unit matrix.

By the well known theorem on the distribution of quadratic forms [RAO 02] the limit distribution of the statistic $\mathbf{Z}_n^T \boldsymbol{\Sigma}^- \mathbf{Z}_n$ is chi-squared with $Tr(\boldsymbol{\Sigma}^- \boldsymbol{\Sigma})$ degrees of freedom, where $\boldsymbol{\Sigma}^-$ is the generalized inverse of $\boldsymbol{\Sigma}$.

Note that

$$\boldsymbol{\Sigma}^- = \mathbf{E}_k + \mathbf{q}\mathbf{q}^T, \quad Tr(\boldsymbol{\Sigma}^- \boldsymbol{\Sigma}) = Tr(\mathbf{E}_k - \mathbf{q}\mathbf{q}^T) = k - 1.$$

So, using the equality $\mathbf{Z}_n^T \mathbf{q} = 0$, we have

$$X_n^2 = \mathbf{Z}_n^T \boldsymbol{\Sigma}^- \mathbf{Z}_n = ||\boldsymbol{Z}_n||^2 \xrightarrow{d} ||\boldsymbol{Z}||^2 \sim \chi^2(k - 1)$$
[2.7]

$\triangle$

The theorem implies the following.

**Pearson's chi-squared test:** the hypothesis $H_0'$ is rejected with an asymptotic significance level $\alpha$ if

$$X_n^2 > \chi_\alpha^2(k-1) \tag{2.8}$$

The hypothesis $H_0'$ can also be verified using the equivalent likelihood ratio test, based on the statistic

$$\Lambda_n = \frac{\sup_{\pi=\pi_0} L(\pi)}{\sup_\pi L(\pi)} = \frac{L(\pi_0)}{L(\hat{\pi})} = n^n \prod_{i=1}^k \left(\frac{\pi_{i0}}{U_i}\right)^{U_i}$$

Under the hypothesis $H_0'$ (see Appendix A, comment A3)

$$R_n = -2\ln\Lambda_n = 2\sum_{i=1}^k U_i \ln\frac{U_i}{n\pi_{i0}} \xrightarrow{d} V \sim \chi^2(k-1), \quad \text{as} \quad n\to\infty \tag{2.9}$$

So the statistics $R_n$ and $X_n^2$ are asymptotically equivalent.

**Likelihood ratio test:** hypothesis $H_0'$ is rejected with an asymptotic significance level $\alpha$ if

$$R_n > \chi_\alpha^2(k-1) \tag{2.10}$$

**Comment 2.1.** Both given tests are not exact, they are used when the sample size $n$ is large. The accuracy of the conclusions depends on the accuracy of the approximations [2.7, 2.9]. If the number of intervals is too large then $U_i$ tend to take values 0 or 1 for all $i$ and the approximation is poor. So the number of intervals $k$ must not be too large.

*Rule of thumb:* choose the grouping intervals to obtain $n\pi_{i0} \geq 5$.

**Comment 2.2.** If the accuracy of the approximation of the distribution of the statistic $X_n^2$ (or $R_n$) is suspected not to be sufficient then the $P$-value can be computed by simulation.

Suppose that the realization of the statistic $X_n^2$ is $x_n^2$. $N$ values of the random vector $U \sim \mathcal{P}(n, \pi_0)$ are simulated and for each value of $U$ the corresponding value of the statistics $X_n^2$ can be computed. Suppose that $M$ is the number of values greater then $x_n^2$. Then the $P$-value is approximated by $M/N$. The hypothesis $H_0'$ is rejected with an approximated significance level $\alpha$ if $M/N < \alpha$. The accuracy of this test depends on the number of simulations $N$.

**Comment 2.3.** If the hypothesis $H_0'$ is rejected then the hypothesis $H_0$ is also rejected because it is narrower. If the data do not contradict the hypothesis $H_0'$ then we have no reason to reject the hypothesis $H_0$.

**Comment 2.4.** If the distribution of the random variable $X$ is discreet and concentrated at the points $x_1, ..., x_k$ then grouping is not needed in such a case. $U_i$ is the observed number of the value $x_i$.

**Comment 2.5.** As was noted, the hypotheses $H_0$ and $H_0'$ are not equivalent in general. Hypothesis $H_0'$ states that the increment of the cumulative distribution function in the $j$-th interval is $\pi_{j0}$ but the behavior of the cumulative distribution function inside the interval is not specified. If $n$ is large then the number of grouping intervals can be increased and thus the hypotheses become closer.

**Comment 2.6.** If the hypothesis $H_0'$ does not hold and $U \sim \mathcal{P}_k(n, \pi)$ then the distributions of the statistics $R_n$ and $X_n^2$ are approximately non-central chi-squared with $k - 1$ degrees of freedom and *the parameter of non-centrality*

$$\Delta = 2n \sum_{j=1}^{k} \pi_j \ln \frac{\pi_j}{\pi_{j0}} \approx \delta = n \sum_{i=1}^{k} \frac{(\pi_j - \pi_{j0})^2}{\pi_{j0}}. \qquad [2.11]$$

**Example 2.1.** Random number generator generated $n = 80$ random numbers. Ordered values are given in the following table:

0.0100 0.0150 0.0155 0.0310 0.0419 0.0456 0.0880 0.1200
0.1229 0.1279 0.1444 0.1456 0.1621 0.1672 0.1809 0.1855
0.1882 0.1917 0.2277 0.2442 0.2456 0.2476 0.2538 0.2552
0.2681 0.3041 0.3128 0.3810 0.3832 0.3969 0.4050 0.4182
0.4259 0.4365 0.4378 0.4434 0.4482 0.4515 0.4628 0.4637
0.4668 0.4773 0.4799 0.5100 0.5309 0.5391 0.6033 0.6283
0.6468 0.6519 0.6686 0.6689 0.6865 0.6961 0.7058 0.7305
0.7337 0.7339 0.7440 0.7485 0.7516 0.7607 0.7679 0.7765
0.7846 0.8153 0.8445 0.8654 0.8700 0.8732 0.8847 0.8935
0.8987 0.9070 0.9284 0.9308 0.9464 0.9658 0.9728 0.9872

Verify the hypothesis $H_0$ that a realization from the uniform distribution $U(0, 1)$ was observed.

Divide the region of possible values $(0, 1)$ into $k = 5$ intervals of equal length:

$$(0; 0.2), [0.2; 0.4), ..., [0.8; 1)$$

We have: $U_j = 18, 12.16, 19, 15$. Let us consider the wider hypothesis $H_0' : \pi_i = 0.2, \ i = 1, ..., 5$. We obtain

$$X_n^2 = \frac{1}{n} \sum_{i=1}^{k} \frac{U_i^2}{\pi_{i0}} - n = \frac{1}{80} \frac{18^2 + 12^2 + 16^2 + 19^2 + 15^2}{0.2} - 80 = 1.875$$

The asymptotic $P$-value is $pv_a = \mathbf{P}\{\chi_4^2 > 1.875) = 0.7587$. The data do not contradict the hypothesis $H_0'$. So we have no basis for rejecting the hypothesis $H_0$. The likelihood ratio test gives the same result because the value of the statistic $R_n$ is 1.93 and $pv_a = \mathbf{P}\{\chi_4^2 > 1.93\} = 0.7486$.

**Comment 2.7.** Tests [2.8] and [2.10] are used not only for the simple hypothesis $H_0$ but also directly for the hypothesis $H_0'$ (see the following example).

**Example 2.2.** Over a long period it has been established that the proportion of the first and second quality units produced in the factory are 0.35 and 0.6, respectively, and the proportion of defective units is 0.05. In a quality inspection, 300 units were checked and 115, 165 and 20 units of the above-considered qualities were found. Did the quality of the product remain the same?

In this example $U_1 = 115$, $U_2 = 165$, $U_3 = 20$, $n = 300$. The zero hypothesis is

$$H_0' : \pi_1 = 0.35, \ \pi_2 = 0.60, \ \pi_3 = 0.05$$

The values of the statistics [2.9] and [2.5] are

$$R_n = 3.717, \quad X_n^2 = 3.869$$

The number of degrees of freedom is $k - 1 = 3 - 1 = 2$.

The $P$-values corresponding to the likelihood ratio and Pearson's chi-squared tests are

$$pv_1 = \mathbf{P}\{\chi_2^2 > 3.717\} = 0.1559$$

$$\text{and} \quad pv_2 = \mathbf{P}\{\chi_2^2 > 3.869\} = 0.1445$$

respectively. The data do not contradict the zero hypothesis.

## 2.3. Pearson's    goodness-of-fit    test:    composite hypothesis

Suppose that $\mathbf{X} = (X_1, ..., X_n)^T$ is a simple sample obtained by observing a random variable $X$ with a cdf from the family $\mathcal{P} = \{F : F \in \mathcal{F}\}$.

### Composite hypothesis

$$H_0 : F(x) \in \mathcal{F}_0 = \{F_0(x; \boldsymbol{\theta}), \boldsymbol{\theta} \in \Theta\} \subset \mathcal{F} \qquad [2.12]$$

meaning that the cdf $F$ belongs to the cdf class $\mathcal{F}_0$ of form $F_0(x; \boldsymbol{\theta})$; here $\boldsymbol{\theta} = (\theta_1, ..., \theta_s)^T \in \Theta \subset \mathbf{R}^s$ is an unknown $s$-dimensional parameter and $F_0$ is a specified cumulative distribution function.

For example, the hypothesis may mean that the probability distribution of $X$ belongs to the family of normal, exponential, Poisson, binomial or other distributions.

As in the previous section, divide the abscissas axis into $k > s + 1$ smaller intervals $(a_{i-1}, a_i]$ and denote by $U_j$ the number of observations belonging to the $j$-th interval, $j = 1, 2...., k$.

The grouped sample $\boldsymbol{U} = (U_1, ..., U_k)^T$ has a $k$-dimensional multinomial distribution $\mathcal{P}_k(n, \boldsymbol{\pi})$; here

$$\boldsymbol{\pi} = (\pi_1, ..., \pi_k)^T$$

$$\pi_i = \mathbf{P}\{X \in (a_i - 1, a_i]\} = F(a_i) - F(a_{i-1}), \quad F \in \mathcal{F}$$

If the hypothesis $H_0$ is true then the wider hypothesis

$$H'_0 : \boldsymbol{\pi} = \boldsymbol{\pi}(\boldsymbol{\theta}), \quad \boldsymbol{\theta} \in \Theta$$

also holds; here

$$\boldsymbol{\pi}(\boldsymbol{\theta}) = (\pi_1(\boldsymbol{\theta}), ..., \pi_k(\boldsymbol{\theta}))^T, \quad \pi_i(\boldsymbol{\theta}) = F_0(a_i; \boldsymbol{\theta}) - F_0(a_{i-1}; \boldsymbol{\theta})$$
$$[2.13]$$

The last hypothesis means that the parameters of the multinomial random vector $U$ can be expressed as specified functions [2.13] of the parameters $\theta_1, ..., \theta_s;\ s+1 < k$.

The Pearson's chi-squared statistic (see [2.5])

$$X_n^2(\boldsymbol{\theta}) = \sum_{i=1}^{k} \frac{(U_i - n\pi_i(\boldsymbol{\theta}))^2}{n\pi_i(\boldsymbol{\theta})} = \frac{1}{n}\sum_{i=1}^{k} \frac{U_i^2}{\pi_i(\boldsymbol{\theta})} - n \qquad [2.14]$$

cannot be computed because the parameter $\theta$ is unknown. It is natural to replace the unknown parameters in the expression [2.14] by their estimators and to investigate the properties of the obtained statistic. If the maximum likelihood estimator of the parameter $\theta$ obtained from the initial *non-grouped* data is used then the limit distribution depends on the distribution $F(x; \boldsymbol{\theta})$ (so on the parameter $\theta$).

We shall see that if *grouped* data are used then estimators of the parameter $\theta$ can be found such that the limit distribution of the obtained statistic is chi-squared with $k - s - 1$ degrees of freedom, so does not depend on $\theta$.

Let us consider several examples of such estimators.

1) Under the hypothesis $H_0$, the likelihood function from the data $U$ and its logarithm are

$$\tilde{L}(\boldsymbol{\theta}) = \frac{n!}{U_1!\ldots U_k!} \prod_{i=1}^{k} \pi_i^{U_i}(\boldsymbol{\theta}), \quad \tilde{\ell}(\boldsymbol{\theta}) = \sum_{i=1}^{k} U_i \ln \pi_i(\boldsymbol{\theta}) + C \ [2.15]$$

so the *maximum likelihood estimator* $\theta_n^*$ of the parameter $\theta$ from the grouped data verifies the system of equations

$$\frac{\partial \tilde{\ell}(\boldsymbol{\theta})}{\partial \theta_j} = \sum_{i=1}^{k} \frac{U_i}{\pi_i(\boldsymbol{\theta})} \frac{\partial \pi_i(\boldsymbol{\theta})}{\partial \theta_j} = 0, \quad j = 1, 2....,s \qquad [2.16]$$

Let us define the Pearson statistic obtained from [2.14]. Replacing $\theta$ by $\theta_n^*$ in [2.14] we obtain the following statistic:

$$X_n^2(\theta_n^*) = \sum_{i=1}^{k} \frac{(U_i - n\pi_i(\theta_n^*))^2}{n\pi_i(\theta_n^*)} \qquad [2.17]$$

2) Another estimator $\tilde{\theta}_n$ of the parameter $\theta$, called the *minimum chi-squared estimator*, is obtained by minimizing [2.14] with respect to $\theta$

$$X_n^2(\tilde{\theta}_n) = \inf_{\theta \in \Theta} X_n^2(\theta) = \inf_{\theta \in \Theta} \sum_{i=1}^{k} \frac{(U_i - n\pi_i(\theta))^2}{n\pi_i(\theta)} \qquad [2.18]$$

3) To find the estimator $\tilde{\theta}_n$, complicated systems of equations must be solved, so statistic [2.14] is often modified by replacing the denominator by $U_i$. This method is called the *modified chi-squared minimum* method. The estimator $\bar{\theta}_n$ obtained by this method is found from the condition

$$X_n^2(\bar{\theta}_n) = \inf_{\theta \in \Theta} \sum_{i=1}^{k} \frac{(U_i - n\pi_i(\theta))^2}{U_i} \qquad [2.19]$$

Besides these three chi-squared type statistics [2.17–2.19] we may use the following.

4) The *likelihood ratio statistic* obtained from the grouped data

$$R_n = -2\ln \frac{\sup_{\theta \in \Theta} \tilde{L}(\theta)}{\sup_\pi L(\pi)} = -2\ln \frac{\sup_{\theta \in \Theta} \prod_{i=1}^{k} \pi_i^{U_i}(\theta)}{\sup_\pi \prod_{i=1}^{k} \pi_i^{U_i}}$$

$$= 2\sum_{i=1}^{k} U_i \ln \frac{U_i}{n\pi_i(\theta_n^*)}$$

This statistic can be written in the form

$$R_n = R_n(\boldsymbol{\theta}_n^*) = \inf_{\boldsymbol{\theta} \in \Theta} R_n(\boldsymbol{\theta}), \quad R_n(\boldsymbol{\theta}) = 2\sum_{i=1}^{k} U_i \ln \frac{U_i}{n\pi_i(\boldsymbol{\theta})} \quad [2.20]$$

We shall show that the statistics $X^2(\tilde{\boldsymbol{\theta}}_n)$, $X^2(\bar{\boldsymbol{\theta}}_n)$, $X^2(\boldsymbol{\theta}_n^*)$ and $R_n(\boldsymbol{\theta}_n^*)$ are asymptotically equivalent as $n \to \infty$.

Suppose that $\{Y_n\}$ is any sequence of random variables. We write $Y_n = o_P(1)$, if $Y_n \xrightarrow{P} 0$, and we write $Y_n = O_P(1)$, if

$$\forall \varepsilon > 0 \quad \exists c > 0 : \sup_n \mathbf{P}\{|Y_n| > c\} < \varepsilon$$

Prokhorov's theorem [VAN 00] implies that if $\exists Y : Y_n \xrightarrow{d} Y$, $n \to \infty$, then $Y_n = O_P(1)$.

## Conditions A

1) For all $i = 1, \ldots, k$ and all $\boldsymbol{\theta} \in \Theta$

$$0 < \pi_i(\boldsymbol{\theta}) < 1, \quad \pi_1(\boldsymbol{\theta}) + \cdots + \pi_k(\boldsymbol{\theta}) = 1$$

2) The functions $\pi_i(\boldsymbol{\theta})$ have continuous first- and second-order partial derivatives on the set $\Theta$.

3) The rank of the matrix

$$\mathbf{B} = \left[\frac{\partial \pi_i(\boldsymbol{\theta})}{\partial \theta_j}\right]_{k \times s}, \quad i = 1, \ldots, k, \quad j = 1, \ldots, s,$$

is $s$.

**Lemma 2.1.** *Suppose that the hypothesis $H_0$ holds. Then under conditions A the estimators $\pi_i(\tilde{\boldsymbol{\theta}}_n)$, $\pi_i(\bar{\boldsymbol{\theta}}_n)$ and $\pi_i(\boldsymbol{\theta}_n^*)$ are $\sqrt{n}$-consistent, i.e. $\sqrt{n}(\tilde{\pi}_{in} - \pi_i) = O_P(1)$.*

**Proof.** For brevity we shall not write the argument of the functions $\pi_i(\boldsymbol{\theta})$.

Let us consider the estimator $\tilde{\pi}_{in}$. Since $0 \leq \tilde{\pi}_{in} \leq 1$, we have $\tilde{\pi}_{in} = O_P(1)$. Since $U_i/n \overset{P}{\to} \pi_{i,}$, the inequalities (we use the definition of the chi-squared minimum estimator)

$$\sum_{i=1}^{k} (U_i/n - \tilde{\pi}_{in})^2 \leq \sum_{i=1}^{k} \frac{(U_i/n - \tilde{\pi}_{in})^2}{\tilde{\pi}_{in}} \leq \sum_{i=1}^{k} \frac{(U_i/n - \pi_i)^2}{\pi_i} = o_P(1)$$

imply that for all $i$: $U_i/n - \tilde{\pi}_{in} = o_P(1)$ and

$$\tilde{\pi}_{in} - \pi_i = (\tilde{\pi}_{in} - U_i/n) + (U_i/n - \pi_i) = o_P(1)$$

Since $\sqrt{n}(\hat{\pi}_i - \pi_i) \overset{d}{\to} Z_i \sim N(0, \pi_i(1 - \pi_i))$, we have $(U_i - n\pi_i)/\sqrt{n} = \sqrt{n}(\hat{\pi}_i - \pi_i) = O_P(1)$. So from the inequality

$$\sum_{i=1}^{k} \frac{(U_i - n\tilde{\pi}_{in})^2}{n} \leq \sum_{i=1}^{k} \frac{(U_i - n\tilde{\pi}_{in})^2}{n\tilde{\pi}_{in}} \leq \sum_{i=1}^{k} \frac{(U_i - n\pi_i)^2}{n\pi_i} = O_P(1)$$

we have that for all $i$: $(U_i - n\tilde{\pi}_{in})/\sqrt{n} = O_P(1)$, and

$$\sqrt{n}(\tilde{\pi}_{in} - \pi_i) = \frac{n\tilde{\pi}_{in} - n\pi_i}{\sqrt{n}} = \frac{n\tilde{\pi}_{in} - U_i}{\sqrt{n}} + \frac{U_i - n\pi_i}{\sqrt{n}} = O_P(1)$$

Analogously we obtain

$$\sum_{i=1}^{k} \frac{(U_i - n\bar{\pi}_{in})^2}{U_i} \leq \sum_{i=1}^{k} \frac{(U_i - n\pi_i)^2}{U_i} \leq$$

$$\sum_{i=1}^{k} \frac{(\sqrt{n}(U_i/n - \pi_i))^2}{U_i/n} = O_P(1)$$

and

$$\sqrt{n}(\bar{\pi}_{in} - \pi_i) = \frac{n\bar{\pi}_{in} - U_i}{\sqrt{n}} + \frac{U_i - n\pi_i}{\sqrt{n}} = O_P(1)$$

Let us consider the estimator $\pi_{in}^*$. $\boldsymbol{\theta}_n^*$ is the ML estimator so under the conditions of the theorem the sequence $\sqrt{n}(\boldsymbol{\theta}_n^* - \boldsymbol{\theta})$ has the limit normal distribution. By the delta method [VAN 00]

$$\sqrt{n}(\pi_{in}^* - \pi_i) = \sqrt{n}(\pi_i(\boldsymbol{\theta}_n^*) - \pi_i(\boldsymbol{\theta})) =$$

$$\dot{\pi}_i^T(\boldsymbol{\theta})\sqrt{n}(\boldsymbol{\theta}_n^* - \boldsymbol{\theta}) + o_P(1) = O_P(1)$$

$\triangle$

**Theorem 2.2.** *Under conditions A the statistics* $X^2(\tilde{\boldsymbol{\theta}}_n)$, $X^2(\bar{\boldsymbol{\theta}}_n)$, $X^2(\boldsymbol{\theta}_n^*)$ *and* $R_n(\boldsymbol{\theta}_n^*)$ *are asymptotically equivalent as* $n \to \infty$:

$$X^2(\tilde{\boldsymbol{\theta}}_n) = X^2(\bar{\boldsymbol{\theta}}_n) + o_P(1) = X^2(\boldsymbol{\theta}_n^*) + o_P(1) = R_n(\boldsymbol{\theta}_n^*) + o_P(1)$$

*The distribution of each statistic converges to the chi-squared distribution with* $k - s - 1$ *degrees of freedom.*

**Proof.** Suppose that $\hat{\boldsymbol{\theta}}$ is an estimator of $\boldsymbol{\theta}$ such that

$$\hat{\boldsymbol{\pi}}_n = (\hat{\pi}_{1n}, \ldots, \hat{\pi}_{kn})^T = (\pi_1(\hat{\boldsymbol{\theta}}), \ldots, \pi_k(\hat{\boldsymbol{\theta}}))^T$$

is the $\sqrt{n}$-consistent estimator of the parameter $\pi$. From the definition of $\sqrt{n}$-consistency and the convergence $U_i/n \xrightarrow{P} \pi_i$ we have that for all $i$

$$\hat{\pi}_{in} - \frac{U_i}{n} = o_P(1), \quad \sqrt{n}\left(\hat{\pi}_{in} - \frac{U_i}{n}\right) = O_P(1), \quad \frac{U_i}{n} = \pi_i + o_P(1)$$

Using the last inequalities, the Taylor expansion

$$\ln(1 + x) = x - x^2/2 + o(x^2), \quad x \to 0$$

and the equality $U_1 + \ldots + U_k = n$, we obtain

$$\frac{1}{2}R_n(\hat{\boldsymbol{\theta}}_n) = \sum_{i=1}^{k} U_i \ln \frac{U_i}{n\hat{\pi}_i} = -\sum_{i=1}^{k} U_i \ln\left(1 + \frac{n\hat{\pi}_i}{U_i} - 1\right) =$$

$$-\sum_{i=1}^{k} U_i \ln\left(1 + \frac{\hat{\pi}_i - U_i/n}{U_i/n}\right) = -\sum_{i=1}^{k} U_i \left(\frac{\hat{\pi}_i - U_i/n}{U_i/n}\right) +$$

$$\frac{1}{2}\sum_{i=1}^{k} U_i \left(\frac{\hat{\pi}_i - U_i/n}{U_i/n}\right)^2 + \sum_{i=1}^{k} U_i o_P\left(\left(\frac{\hat{\pi}_i - U_i/n}{U_i/n}\right)^2\right) =$$

$$-n\sum_{i=1}^{k} \hat{\pi}_i + \sum_{i=1}^{k} U_i + \frac{1}{2}\sum_{i=1}^{k} \frac{(U_i - n\hat{\pi}_i)^2}{U_i} + o_P(1) =$$

$$\frac{1}{2}\sum_{i=1}^{k} \frac{(U_i - n\hat{\pi}_i)^2}{U_i} + o_P(1) =$$

$$\frac{1}{2}\sum_{i=1}^{k} \frac{(U_i - n\hat{\pi}_i)^2}{n\hat{\pi}_i} - \frac{1}{2}\sum_{i=1}^{k} \frac{(U_i - n\hat{\pi}_i)^3}{U_i n\hat{\pi}_i} + o_P(1) =$$

$$\frac{1}{2}\sum_{i=1}^{k} \frac{(U_i - n\hat{\pi}_i)^2}{n\hat{\pi}_i} + o_P(1) = \frac{1}{2}X_n^2(\hat{\boldsymbol{\theta}}_n) + o_P(1)$$

Taking $\hat{\boldsymbol{\theta}}_n = \boldsymbol{\theta}_n^*$ and $\hat{\boldsymbol{\theta}}_n = \tilde{\boldsymbol{\theta}}_n$, the last equalities imply

$$X_n^2(\boldsymbol{\theta}_n^*) = R_n(\boldsymbol{\theta}_n^*) + o_P(1), \quad X_n^2(\tilde{\boldsymbol{\theta}}_n) = R_n(\tilde{\boldsymbol{\theta}}_n) + o_P(1)$$

From the definition of $\tilde{\boldsymbol{\theta}}_n$ we obtain $X_n^2(\tilde{\boldsymbol{\theta}}_n) \leq X_n^2(\boldsymbol{\theta}_n^*)$, and from definition [2.20] of $R_n$ we obtain $R_n(\boldsymbol{\theta}_n^*) \leq R_n(\tilde{\boldsymbol{\theta}}_n)$. So

$$X_n^2(\tilde{\boldsymbol{\theta}}_n) \leq X_n^2(\boldsymbol{\theta}_n^*) = R_n(\boldsymbol{\theta}_n^*) + o_P(1) \leq$$

$$R_n(\tilde{\boldsymbol{\theta}}_n) + o_P(1) = X_n^2(\tilde{\boldsymbol{\theta}}_n) + o_P(1)$$

These inequalities imply

$$X_n^2(\tilde{\boldsymbol{\theta}}_n) = R_n(\boldsymbol{\theta}_n^*) + o_P(1)$$

Analogously

$$X_n^2(\bar{\boldsymbol{\theta}}_n) = R_n(\boldsymbol{\theta}_n^*) + o_P(1)$$

The $(k-1)$-dimensional vector $(\pi_1, \ldots, \pi_{k-1})^T$ is a function of the $s$-dimensional parameter $\theta$, so the limit distribution of the likelihood ratio statistic $R_n(\boldsymbol{\theta}_n^*)$ is chi-squared with $k-s-1$ degrees of freedom (see Appendix A, comment A.4). The same limit distributions have other considered statistics.

$\triangle$

The theorem implies the following asymptotic tests.

**Chi-squared test:** hypothesis $H_0'$ is rejected with an asymptotic significance level $\alpha$ if

$$X^2(\hat{\boldsymbol{\theta}}_n) > \chi_\alpha^2(k - 1 - s) \qquad [2.21]$$

here $\hat{\boldsymbol{\theta}}_n$ is any of the estimators $\tilde{\boldsymbol{\theta}}_n, \boldsymbol{\theta}_n^*, \bar{\boldsymbol{\theta}}_n$.

**Likelihood ratio test:** hypothesis $H_0'$ is rejected with an asymptotic significance level $\alpha$ if

$$R_n(\boldsymbol{\theta}_n^*) > \chi_\alpha^2(k - 1 - s) \qquad [2.22]$$

If hypothesis $H_0'$ is rejected then the hypothesis $H_0$ is also rejected.

**Example 2.3.** In reliability testing the numbers of failed units $U_i$ in time intervals $[a_{i-1}, a_i)$, $i = 1, ..., 11$, were fixed.

The data are given in the following table.

| $i$ | $(a_{i-1}, a_i]$ | $U_i$ | $i$ | $(a_{i-1}, a_i]$ | $U_i$ |
|---|---|---|---|---|---|
| 1 | (0,100] | 8 | 7 | (600,700 ] | 25 |
| 2 | (100,200] | 12 | 8 | (700,800] | 18 |
| 3 | (200,300] | 19 | 9 | (800,900] | 15 |
| 4 | (300,400] | 23 | 10 | (900,1000] | 14 |
| 5 | (400,500] | 29 | 11 | (1000,$\infty$) | 18 |
| 6 | (500,600] | 30 | | | |

Verify the hypothesis stating that the failure times have the Weibull distribution.

By [2.20], the estimator $(\theta_n^*, \nu_n^*)$ minimizes the function

$$R_n(\theta, \nu) = 2 \sum_{i=1}^{k} U_i \ln \frac{U_i}{n \pi_i(\theta, \nu)}, \quad \pi_i(\theta, \nu) = e^{-(a_{i-1}/\theta)^\nu} - e^{-(a_i/\theta)^\nu}$$

By differentiating this function with respect to the parameters and equating the partial derivatives to zero, the following system of equations is obtained for the estimators $\theta_n^*$ and $\nu_n^*$

$$\sum_{i=1}^{k} U_i \frac{a_{i-1}^\nu e^{-(a_{i-1}/\theta)^\nu} - a_i^\nu e^{-(a_i/\theta)^\nu}}{e^{-(a_{i-1}/\theta)^\nu} - e^{-(a_i/\theta)^\nu}} = 0$$

$$\sum_{i=1}^{k} U_i \frac{a_{i-1}^\nu e^{-(a_{i-1}/\theta)^\nu} \ln a_{i-1} - a_i^\nu e^{-(a_i/\theta)^\nu} \ln a_i}{e^{-(a_{i-1}/\theta)^\nu} - e^{-(a_i/\theta)^\nu}} = 0$$

By solving this system of equations or directly minimizing the function $R_n(\theta, \nu)$, we obtain the estimators $\theta^* = 649.516$ and $\nu^* = 2.004$ of the parameters $\theta$ and $\nu$. Minimizing the right-hand sides of [2.18] and [2.19] we obtain the values of the estimators $\tilde{\theta} = 647.380$, $\tilde{\nu} = 1.979$ and $\bar{\theta} = 653.675$, $\bar{\nu} = 2.052$. Using the obtained estimators, we have the following values of the test statistics

$$R_n(\theta^*, \nu^*) = 4.047, \quad X_n^2(\theta^*, \nu^*) = 4.377, \quad X_n^2(\tilde{\theta}, \tilde{\nu}) = 4.324 \quad \text{and}$$

$$X_n^2(\bar{\theta}, \bar{\nu}) = 3.479$$

The number of degrees of freedom is $k - s - 1 = 8$. Since the $P$-values – 0.853; 0.822; 0.827; 0.901 – are not small, there is no reason to reject the zero hypothesis.

## 2.4. Modified    chi-squared    test    for    composite hypotheses

The classical Pearson's chi-squared test has drawbacks, especially in the case of continuous distributions.

First, in investigating the asymptotic distributions of statistics, the assumption that grouping intervals are chosen independently of data was made in Theorem 2.1. In practice, grouping intervals are chosen depending on the obtained realization of the sample: a minimum five observations must fall into each interval. So, we cannot formally use the results of Theorem 2.1.

Secondly, the chi-squared minimum estimators from the grouped data are not optimal because they do not use all the data and their computation requires the solution of complicated systems of equations.

Thirdly, the asymptotically optimal estimators (in regular models these are the ML estimators) from all data cannot be used in Pearson's statistic because the limit distribution of Pearson's statistic depends on unknown parameters.

Let us consider modifications of Pearson's statistic which do not have the above-mentioned drawbacks.

### 2.4.1. *General case*

Suppose that the composite hypothesis [2.12] holds. Let us consider the limit distribution of the random vector $Z_n = (Z_{1n}, \ldots, Z_{kn})^T$

$$Z_{jn} = \frac{U_j - n\pi_j(\hat{\boldsymbol{\theta}}_n)}{\sqrt{n\pi_j(\hat{\boldsymbol{\theta}}_n)}} = \sqrt{n}\left(\frac{U_j}{n} - \pi_j(\hat{\boldsymbol{\theta}}_n)\right) / \sqrt{\pi_j(\hat{\boldsymbol{\theta}}_n)} \quad [2.23]$$

where $\hat{\boldsymbol{\theta}}_n$ is the maximum likelihood estimator from a non-grouped simple sample $\mathbf{X} = (X_1, \ldots, X_n)^T$. This estimator maximizes the loglikelihood function (see Appendix A)

$$\ell(\boldsymbol{\theta}) = \sum_{i=1}^{n} \ell_i(\boldsymbol{\theta}), \quad \ell_i(\boldsymbol{\theta}) = \ln f(X_i, \boldsymbol{\theta})$$

If the model $\{F_0(x;\boldsymbol{\theta}), \boldsymbol{\theta} \in \Theta\}$ is regular (see Appendix A) then the Fisher information matrix of $X_1$ is $i(\boldsymbol{\theta}) = \mathbf{E}_{\boldsymbol{\theta}}\dot{\ell}_1(\boldsymbol{\theta})\dot{\ell}_1(\boldsymbol{\theta})^T = -\mathbf{E}_{\boldsymbol{\theta}}\ddot{\ell}_1(\boldsymbol{\theta})$, where $\dot{\ell}_1(\boldsymbol{\theta})$ is the vector of the partial derivatives of $\ell_1(\boldsymbol{\theta})$ with respect to the parameters $\theta_1, ..., \theta_s$ and $\ddot{\ell}_1(\boldsymbol{\theta})$ – the matrix of the second partial derivatives.

**Theorem 2.3.** *Suppose that the parametric model $\{F_0(x;\boldsymbol{\theta}), \boldsymbol{\theta} \in \Theta\}$ is regular and conditions A (see Appendix A) are satisfied. Then*

$$\boldsymbol{Z}_n \xrightarrow{d} \boldsymbol{Z} \sim N_k(\mathbf{0}, \boldsymbol{\Sigma}(\boldsymbol{\theta}))$$

$$Y_n^2 = \boldsymbol{Z}_n^T \boldsymbol{\Sigma}^-(\hat{\boldsymbol{\theta}}_n)\boldsymbol{Z}_n \xrightarrow{d} Y^2 \sim \chi^2(k-1) \qquad [2.24]$$

*where*

$$\boldsymbol{\Sigma}(\boldsymbol{\theta}) = (\mathbf{E}_k - \mathbf{q}(\boldsymbol{\theta})\mathbf{q}^T(\boldsymbol{\theta}))(\mathbf{E}_k - C^T(\boldsymbol{\theta})i^{-1}(\boldsymbol{\theta})C(\boldsymbol{\theta}))$$

$$\boldsymbol{\Sigma}^-(\boldsymbol{\theta}) = (\mathbf{E}_k - C^T(\boldsymbol{\theta})i^{-1}(\boldsymbol{\theta})C(\boldsymbol{\theta}))^{-1}$$

*where $\boldsymbol{\Sigma}^-$ is the generalized inverse of the matrix $\boldsymbol{\Sigma}$, i.e. the matrix satisfying the equality $\boldsymbol{\Sigma}\boldsymbol{\Sigma}^-\boldsymbol{\Sigma} = \boldsymbol{\Sigma}$; $\mathbf{E}_k$ is the $k \times k$ unit matrix; $i(\boldsymbol{\theta})$ – the Fisher information matrix of $X_1$ (see Appendix A)*

$$C(\boldsymbol{\theta}) = [c_{ij}(\boldsymbol{\theta})]|_{s \times k} = \left[ \frac{1}{\sqrt{\pi_j(\boldsymbol{\theta})}} \frac{\partial \pi_j(\boldsymbol{\theta})}{\partial \theta_i} \right]_{s \times k}$$

$$\mathbf{q}(\boldsymbol{\theta}) = (\sqrt{\pi_1(\boldsymbol{\theta})}, \ldots, \sqrt{\pi_k(\boldsymbol{\theta})})^T$$

**Proof.** In regular models the ML estimator satisfies the equality (see Appendix A)

$$\sqrt{n}(\hat{\boldsymbol{\theta}}_n - \boldsymbol{\theta}) = i^{-1}(\boldsymbol{\theta})\frac{1}{\sqrt{n}}\dot{\ell}(\boldsymbol{\theta}) + o_P(1) \qquad [2.25]$$

where $\dot{\ell}(\boldsymbol{\theta})$ is the vector of the score functions from a non-grouped sample.

The delta method [VAN 00] implies

$$\sqrt{n}(\pi_j(\hat{\boldsymbol{\theta}}_n)) - \pi_j(\boldsymbol{\theta})) = \dot{\pi}_j^T(\boldsymbol{\theta})\sqrt{n}(\hat{\boldsymbol{\theta}}_n - \boldsymbol{\theta}) + o_P(1)$$

so using the expression [2.23] of $Z_{jn}$, the formula [2.25] and the equality $\pi_j(\hat{\boldsymbol{\theta}}_n) = \pi_j(\boldsymbol{\theta}) + o_P(1)$, we obtain

$$Z_{jn} = \left\{ \sqrt{n} \left( \frac{U_j}{n} - \pi_j(\boldsymbol{\theta}) \right) - \dot{\pi}_j^T(\boldsymbol{\theta}) i^{-1}(\boldsymbol{\theta}) \frac{1}{\sqrt{n}} \dot{\ell}(\boldsymbol{\theta}) \right\} / \sqrt{\pi_j(\boldsymbol{\theta})} + o_P(1)$$

$$=: Y_{jn} + o_P(1)$$

Theorem A1 implies that the limit distributions of the random vectors

$$(\sqrt{n}(U_1/n - \pi_1(\boldsymbol{\theta})), \dots, \sqrt{n}(U_k/n - \pi_k(\boldsymbol{\theta})))^T, \quad \dot{\ell}(\boldsymbol{\theta}) \quad and \quad \boldsymbol{Z}_n$$

are normal. Let us find the mean and the covariance matrix of the random vector $\boldsymbol{Y}_n = (Y_{1n}, \dots, Y_{kn})^T$. Since

$$\mathbf{E}\left( \sqrt{n}(\frac{U_j}{n} - \pi_j(\boldsymbol{\theta}))) \right) = \mathbf{E}(\dot{\ell}(\boldsymbol{\theta})) = 0$$

$$\mathbf{Var}\left( \frac{U_j}{\sqrt{n}} \right) = \pi_j(\boldsymbol{\theta})(1 - \pi_j(\boldsymbol{\theta}))$$

$$\mathbf{Cov}\left( \frac{U_i}{\sqrt{n}}, \frac{U_j}{\sqrt{n}} \right) = -\pi_i(\boldsymbol{\theta})\pi_j(\boldsymbol{\theta}), \quad \mathbf{Var}\left( \frac{1}{\sqrt{n}}\dot{\ell}(\boldsymbol{\theta}) \right) = i(\boldsymbol{\theta})$$

$$\mathbf{Cov}\left( \frac{U_j}{\sqrt{n}}, \frac{1}{\sqrt{n}}\dot{\ell}(\boldsymbol{\theta}) \right) = \mathbf{E}(1_{(a_{j-1}-a_j]}(X_1)\dot{\ell}(\boldsymbol{\theta})) =$$

$$\int_{a_{j-1}}^{a_j} \dot{f}(x, \boldsymbol{\theta})dx = \dot{\pi}_j(\boldsymbol{\theta})$$

we obtain

$$\mathbf{E}(Y_{jn}) = 0, \quad \mathbf{Var}(Y_{jn}) = 1 - \pi_j(\boldsymbol{\theta}) - \dot{\pi}_j^T(\boldsymbol{\theta}) i^{-1}(\boldsymbol{\theta}) \dot{\pi}_j(\boldsymbol{\theta})/\pi_j(\boldsymbol{\theta})$$

and for all $l \neq j$

$$\mathbf{Cov}(Y_{ln}, Y_{jn}) = -\sqrt{\pi_l(\boldsymbol{\theta})\pi_j(\boldsymbol{\theta})} - \dot{\pi}_l^T(\boldsymbol{\theta})i^{-1}(\boldsymbol{\theta})\dot{\pi}_j(\boldsymbol{\theta})/\sqrt{\pi_l(\boldsymbol{\theta})\pi_j(\boldsymbol{\theta})}$$

We obtained that $Z_n \overset{d}{\to} Z \sim N_k(0, \Sigma(\boldsymbol{\theta}))$, where the covariance matrix (we skip the argument $\boldsymbol{\theta}$)

$$\Sigma = \mathbf{E}_k - \mathbf{q}\mathbf{q}^T - C^T i^{-1} C$$

Note that

$$C\mathbf{q} = (\frac{\partial}{\partial\theta_1}\sum_{j=1}^{k}\pi_j(\boldsymbol{\theta}), \ldots, \frac{\partial}{\partial\theta_s}\sum_{j=1}^{k}\pi_j(\boldsymbol{\theta}))^T = (0, \ldots, 0)^T = \mathbf{0}_s$$

So

$$\Sigma = (\mathbf{E}_k - \mathbf{q}\mathbf{q}^T)(\mathbf{E}_k - C^T i^{-1} C)$$

The rank of the matrix $\Sigma$ is $k - 1$ (see exercise 2.4) and its generalized inverse matrix is

$$\Sigma^- = (\mathbf{E}_k - C^T i^{-1} C)^{-1}$$

because the equality

$$(\mathbf{E}_k - \mathbf{q}\mathbf{q}^T)(\mathbf{E}_k - \mathbf{q}\mathbf{q}^T) = (\mathbf{E}_k - \mathbf{q}\mathbf{q}^T)$$

implies the equality $\Sigma\Sigma^-\Sigma = \Sigma$.

The second proposition of the theorem is implied by the following well-known theorem: if

$$Y \sim N_k(0, \Sigma) \quad \text{then} \quad Y^T \Sigma^- Y \sim \chi^2(r)$$

where $r = rank(\Sigma)$.

$\triangle$

**Comment 2.8.** Note that if the matrix $G = i - CC^T$ is non-degenerate (which is true for the probability distributions used in practice) then the generalized inverse of the matrix $\Sigma$ is

$$\Sigma^- = \mathbf{E}_k + C^T G^{-1} C$$

So we do not need to invert the $k \times k$ matrix $\Sigma$ but only the $s \times s$ matrix $G$ (usually $s = 1$ or $s = 2$).

**Proof.**

$$(\mathbf{E}_k - C^T i^{-1} C)(\mathbf{E}_k + C^T G^{-1} C) = \mathbf{E}_k + C^T G^{-1} C - C^T i^{-1} C -$$

$$C^T i^{-1} C C^T G^{-1} C = \mathbf{E}_k + C^T i^{-1} (i G^{-1} - \mathbf{E}_s - C C^T G^{-1}) C =$$

$$\mathbf{E}_k + C^T i^{-1} (G G^{-1} - \mathbf{E}_s) C = \mathbf{E}_k$$

$\triangle$

The test statistic is

$$Y_n^2 = Z_n^T \Sigma^- (\hat{\boldsymbol{\theta}}) Z_n$$

Comment 2.8 implies that this statistic can be written in the form

$$Y_n^2 = X_n^2 + \frac{1}{n} \boldsymbol{v}^T G^{-1} \boldsymbol{v} \qquad [2.26]$$

where

$$X_n^2 = \sum_{i=1}^{k} \frac{(U_i - n\pi_i(\hat{\boldsymbol{\theta}}_n))^2}{n\pi_i(\hat{\boldsymbol{\theta}}_n)} = \sum_{i=1}^{k} \frac{U_i^2}{n\pi_i(\hat{\boldsymbol{\theta}}_n)} - n \qquad [2.27]$$

$$\boldsymbol{v} = (v_1, ..., v_s)^T, \quad v_j = \sum_{i=1}^{k} \frac{U_i}{\pi_i(\hat{\boldsymbol{\theta}}_n)} \frac{\partial \pi_i(\hat{\boldsymbol{\theta}}_n)}{\partial \theta_j}$$

$$G = [g_{rr'}]_{s \times s}, \quad g_{rr'} = i_{rr'} - \sum_{l=1}^{k} \frac{1}{\pi_l(\hat{\boldsymbol{\theta}}_n)} \frac{\partial \pi_l(\hat{\boldsymbol{\theta}}_n)}{\partial \theta_r} \frac{\partial \pi_l(\hat{\boldsymbol{\theta}}_n)}{\partial \theta_{r'}}$$

where $i_{rr'}$ are elements of the matrix $i(\hat{\boldsymbol{\theta}}_n)$.

**Proof.**

$$Y_n^2 = Z_n^T(\mathbf{E}_k + C^T G^{-1} C) Z_n = Z_n^T Z_n +$$
$$\tilde{U}^T C^T G^{-1} C \tilde{U} + n \mathbf{q}^T C^T G^{-1} C \mathbf{q}$$

where

$$\tilde{U} = \left( U_1 / \sqrt{n\pi_1(\hat{\boldsymbol{\theta}}_n)}, ..., U_k / \sqrt{n\pi_k(\hat{\boldsymbol{\theta}}_n)} \right)^T$$

The first term is $Z_n^T Z_n = X_n^2$, the third is 0, and the second is

$$\tilde{U}^T C^T G^{-1} C \tilde{U} = \frac{1}{n} v^T G^{-1} v$$

$\triangle$

**Comment 2.9.** If weak regularity assumptions are satisfied then the limit distribution of the statistics $Y_n^2$ is the same [NIK 73a, NIK 73c] when the limits $a_i$ of the grouping intervals are not pre-fixed but are chosen as data functions in the following way.

Let us fix $k$ positive numbers $p_1, ..., p_k$ satisfying the equality $p_1 + \cdots + p_k = 1$, usually $p_i = 1/k$. Denote by

$$F_0^{-1}(x, \boldsymbol{\theta}) = \inf\{y : F_0(y, \boldsymbol{\theta}) \geq x\}$$

the inverse of the cdf $F$.

The endpoints of the grouping intervals are defined as statistics

$$a_i = a_i(\hat{\boldsymbol{\theta}}_n) = F_0^{-1}(P_i, \hat{\boldsymbol{\theta}}_n), \quad P_i = p_1 + \cdots + p_i,$$

$$i = 1, 2, ..., k, \quad z_0 = -\infty$$

In formula [2.24] the probabilities $\pi_i(\boldsymbol{\theta})$ are

$$\pi_i(\boldsymbol{\theta}) = F_0(a_i(\boldsymbol{\theta}); \hat{\boldsymbol{\theta}}_n) - F_0(a_{i-1}(\boldsymbol{\theta}); \hat{\boldsymbol{\theta}}_n) = \int_{a_{i-1}(\boldsymbol{\theta})}^{a_i(\boldsymbol{\theta})} f(x; \hat{\boldsymbol{\theta}}_n) dx$$

or

$$c_{ij} = \frac{\partial \pi_i(\hat{\boldsymbol{\theta}}_n)}{\partial \theta_j} = f(a_j(\hat{\boldsymbol{\theta}}_n); \hat{\boldsymbol{\theta}}_n) \frac{\partial a_i(\hat{\boldsymbol{\theta}}_n)}{\partial \theta_j} - f(a_{j-1}(\hat{\boldsymbol{\theta}}_n); \hat{\boldsymbol{\theta}}_n) \frac{\partial a_{i-1}(\hat{\boldsymbol{\theta}}_n)}{\partial \theta_j}$$

Then the test statistic [2.26] is written in the form

$$Y_n^2 = X_n^2 + \frac{1}{n} \boldsymbol{v}^T \boldsymbol{G}^{-1} \boldsymbol{v} \qquad [2.28]$$

where

$$X_n^2 = \sum_{i=1}^{k} \frac{(U_i - np_i)^2}{np_i} = \sum_{i=1}^{k} \frac{U_i^2}{np_i} - n \qquad [2.29]$$

$$\boldsymbol{v} = (v_1, ..., v_s)^T, \quad v_j = \frac{c_{1j}U_1}{p_1} + ... + \frac{c_{kj}U_k}{p_k}$$

$$\boldsymbol{G} = [g_{rr'}]_{s \times s}, \quad g_{rr'} = i_{rr'} - \sum_{l=1}^{k} \frac{c_{lr}c_{lr'}}{p_l}$$

**The Nikulin–Rao–Robson test:** hypothesis $H_0$ is rejected with asymptotic significance level $\alpha$ if

$$Y_n^2 > \chi_\alpha^2(k-1) \qquad [2.30]$$

**Comment 2.10.** When applying the modified chi-squared test, it often suffices to compute the statistic $X_n^2$.

Indeed, if $X_n^2 > \chi_\alpha^2(k-1)$, we should reject the zero hypothesis because in such a case $Y_n^2 \geq X_n^2 > \chi_\alpha^2(k-1)$.

### 2.4.2. *Goodness-of-fit for exponential distributions*

Let us consider the hypothesis

$$H : F \in \{G : G(x; \theta) = 1 - e^{-x/\theta}, \ x \geq 0; \ \theta > 0\}$$

meaning that failure times have an exponential distribution.

Denote by $\hat{\theta} = \bar{X}$ the ML estimator of the parameter $\theta$. Using formulas [2.28] we obtain

$$a_i = \hat{\theta} z_i, \quad z_i = -\ln(1 - P_i), \quad \frac{\partial a_i}{\partial \theta} = z_i \qquad [2.31]$$

$$c_i := c_{i1} = \frac{z_i e^{-z_i} - z_{i-1} e^{-z_{i-1}}}{\hat{\theta}} = \frac{b_i}{\hat{\theta}}, \quad i_{11} = \frac{1}{\hat{\theta}^2}$$

So the test statistic $Y_n^2$ is written in the form

$$Y_n^2 = X_n^2 + Q_n, \quad X_n^2 = \sum_{i=1}^{k} \frac{U_i^2}{np_i} - n, \quad Q_n = \frac{v^2}{n\lambda} \qquad [2.32]$$

where

$$v = \sum_{i=1}^{k} \frac{b_i U_i}{p_i}, \quad \lambda = 1 - \sum_{i=1}^{k} \frac{b_i^2}{p_i}$$

**Example 2.4.** The failure times of $n = 70$ electric bulbs are (in months):

5.017 0.146 6.474 13.291 5.126 8.934 10.971 7.863 5.492
13.930 12.708 7.329 5.408 6.808 0.923 4.679 2.242 4.120
12.080 2.502 16.182 6.592 2.653 4.252 8.609 10.419 2.173
3.321 4.086 11.667 19.474 11.067 11.503 2.284 0.926 2.065
4.703 3.744 5.286 5.497 4.881 0.529 10.397 30.621 5.193 7.901
10.220 16.806 10.672 4.209 5.699 20.952 12.542 7.316 0.272
4.380 9.699 9.466 7.928 13.086 8.871 13.000 16.132 9.950
8.449 8.301 16.127 22.698 4.335

Verify the hypothesis that the failure times have an exponential distribution.

We find $\hat{\theta} = \bar{X} = 8.231$. Let $k = 6$. So $p = 1/6, P_i = i/6, i = 1, ..., 6$. Intermediate results are given in the following table.

| $i$ | 0 | 1 | 2 | 3 | 4 | 5 | 6 |
|-----|---|---|---|---|---|---|---|
| $P_i$ | 0.0000 | 0.1667 | 0.3333 | 0.5000 | 0.6667 | 0.8333 | 1.000 |
| $a_i$ | 0.0000 | 1.5005 | 3.3377 | 5.7049 | 9.0426 | 14.7483 | $\infty$ |
| $z_i$ | 0.0000 | 0.1823 | 0.4055 | 0.6931 | 1.0986 | 1.7918 | $\infty$ |
| $b_i$ | – | 0.1519 | 0.1184 | 0.0762 | 0.0196 | -0.0676 | -0.2986 |
| $U_i$ | – | 5 | 8 | 18 | 13 | 18 | 8 |

We have $v = -1.6259$, $\lambda = 0.1778$, $X_n^2 = 13.1429$, $Q_n = 0.2124$ and $Y_n^2 = 13.3553$. The $P$-value of the test based on the statistic $Y_n^2$ is $pv_a = \mathbf{P}\{\chi_5^2 > 13.3553\} = 0.0203$, so the hypothesis is rejected.

### 2.4.3. *Goodness-of-fit for location-scale and shape-scale families*

Let us consider the hypothesis

$$H_0 : F \in \{G : G(x) = \{G_0((x - \mu)/\sigma)),$$

$$x \in \mathbf{R}, \ -\infty < \mu < +\infty, \ 0 < \sigma < \infty\}$$

where $G_0$ is a known cumulative distribution function. The hypothesis means that the distribution of the random variables $X_i$ is from a specified location and scale family.

The modified chi-squares test for such a family can be used for testing the hypothesis

$$H_0^* : F \in \{G : G(x) = G_0((\frac{x}{\theta})^\nu), \ x > 0, \ 0 < \theta, \nu < \infty\}$$

that the distribution of the random variables $X_i$ is from a specified shape and scale family because this class, by the logarithmic transform $Y_i = \ln X_i$, is transformed to a location-shape family. So hypothesis $H_0^*$ can be verified using the test for hypothesis $H_0$ and the transformed data.

Suppose that hypothesis $H_0$ holds. Denote by $\hat{\theta} = (\hat{\mu}, \hat{\sigma})^T$ the ML estimators of unknown parameters. Using formulas [2.28], we obtain

$$a_i = \hat{\mu} + z_i\hat{\sigma}, \ z_i = G_0^{-1}(P_i), \ \frac{\partial a_i}{\partial \mu} = 1, \ \frac{\partial a_i}{\partial \sigma} = z_i, \ g(x) = G_0'(x)$$

[2.33]

$$c_{i1} = \frac{g(z_i) - g(z_{i-1})}{\hat{\sigma}} = \frac{b_{i1}}{\hat{\sigma}}, \quad c_{i2} = \frac{z_i g(z_i) - z_{i-1}g(z_{i-1})}{\hat{\sigma}} = \frac{b_{i2}}{\hat{\sigma}}$$

Set

$$j_{rs} = \int_{-\infty}^{\infty} x^r \left[\frac{g'(x)}{g(x)}\right]^s g(x)dx, \quad r = 0, 1, 2; \quad s = 1, 2$$

The elements of the estimator of the Fisher information matrix $i(\hat{\theta}_n)$ are

$$i_{11} = \frac{j_{02}}{\hat{\sigma}^2}, \quad i_{12} = \frac{j_{12}}{\hat{\sigma}^2}, \quad i_{22} = \frac{j_{22} + 2j_{11} + 1}{\hat{\sigma}^2}$$

[2.34]

So the statistic $Y_n^2$ is written in the form

$$Y_n^2 = X_n^2 + Q_n, \quad Q_n = \frac{\lambda_1 \alpha^2 - 2\lambda_3 \alpha\beta + \lambda_2 \beta^2}{n(\lambda_1 \lambda_2 - \lambda_3^2)}$$

[2.35]

where

$$\alpha = \sum_{i=1}^{k} \frac{b_{i1} U_i}{p_i}, \quad \beta = \sum_{i=1}^{k} \frac{b_{i2} U_i}{p_i}$$

$$\lambda_1 = j_{02} - \sum_{i=1}^{k} \frac{b_{i1}^2}{p_i}, \quad \lambda_2 = j_{22} + 2j_{11} + 1 - \sum_{i=1}^{k} \frac{b_{i2}^2}{p_i}, \quad \lambda_3 = j_{12} - \sum_{i=1}^{k} \frac{b_{i1} b_{i2}}{p_i}$$

Let us consider the expressions of the quantities in the test statistic for some of the most used families of distributions.

**Normal distribution:** $F(x; \mu, \sigma) = \Phi((x - \mu)/\sigma)$, $\Phi(y) = \frac{1}{\sqrt{2\pi}} \int_{-\infty}^{y} e^{-u^2/2} du$.

$$\hat{\mu} = \bar{X} = \frac{1}{n} \sum_{i=1}^{n} X_i, \quad \hat{\sigma} = s, \quad s^2 = \frac{1}{n} \sum_{i=1}^{n} (X_i - \bar{X})^2$$

$$G(x) = \Phi(x), \quad g(x) = \varphi(x) = \Phi'(x)$$

$$j_{02} = 1, \quad j_{12} = 0, \quad j_{22} + 2j_{11} + 1 = 2 \qquad [2.36]$$

**Lognormal distribution:** $F(x; \theta, \nu) = \Phi(\ln(x/\theta)^\nu) = \Phi((\ln x - \mu)/\sigma)$, $x > 0$. Apply the test for the normal distribution using $\ln X_i$ instead of the initial $X_i$.

**Logistic distribution:** $F(x; \mu, \sigma) = 1 - (1 + e^{-(x-\mu)/\sigma})^{-1}$.

$$G(x) = 1 - (1 + e^{-x})^{-1}, \quad g(x) = \frac{e^{-x}}{(1 + e^{-x})^2} = \frac{e^x}{(1 + e^x)^2}$$

$$j_{02} = \frac{1}{3}, \quad j_{12} = 0, \quad j_{22} + 2j_{11} + 1 = \frac{\pi^2 + 3}{9}$$

The ML estimators $\hat{\mu}$ and $\hat{\sigma}$ maximize the loglikelihood function

$$\ell(\mu, \sigma) = -n \ln \sigma + \sum_{i=1}^{n} \frac{X_i - \mu}{\sigma} - 2 \sum_{i=1}^{n} \ln(1 + e^{\frac{X_i - \mu}{\sigma}})$$

**Loglogistic distribution:** $F(x; \theta, \nu) = 1 - (1 + (\frac{x}{\theta})^\nu)^{-1}$, $x > 0$.

Apply the test for the logistic distribution using $\ln X_i$ instead of the initial $X_i$.

**Extreme values distribution:** $F(x; \mu, \sigma) = 1 - e^{-e^{\frac{x-\mu}{\sigma}}}$.

$$G(x) = 1 - e^{-e^x}, \quad g(x) = e^x e^{-e^x}$$

$$j_{02} = 1, \quad j_{12} = \Gamma'(1) + 1, \quad j_{22} + 2j_{11} + 1 = \Gamma''(1) + 2\Gamma'(1) + 4$$

The ML estimators $\hat{\mu}$ and $\hat{\sigma}$ maximize the loglikelihood function

$$\ell(\mu, \sigma) = -n \ln \sigma - \sum_{i=1}^{n} e^{\frac{X_i - \mu}{\sigma}} + \sum_{i=1}^{n} \frac{X_i - \mu}{\sigma}$$

**Weibull distribution:** $F(x; \theta, \nu) = 1 - e^{-\left(\frac{x}{\theta}\right)^{\nu}}, x > 0$.

Apply the test for the extreme value distribution using $\ln X_i$ instead of the initial $X_i$.

**Cauchy distribution:** $F(x; \mu, \sigma) = \frac{1}{\pi} \left( arctg \frac{x-\mu}{\sigma} + \frac{\pi}{2} \right)$.

$$G(x) = \frac{1}{\pi} \left( arctg\, x + \frac{\pi}{2} \right), \quad g(x) = \frac{1}{\pi(1 + x^2)}$$

$$j_{02} = \frac{1}{2}, \quad j_{12} = 0, \quad j_{22} + 2j_{11} + 1 = \frac{1}{2}$$

The ML estimators $\hat{\mu}$ and $\hat{\sigma}$ maximize the loglikelihood function

$$\ell(\mu, \sigma) = -n \ln \sigma - \sum_{i=1}^{n} \ln(1 + (\frac{X_i - \mu}{\sigma})^2)$$

**Comment 2.12.** From many authors' results it can be concluded that if alternatives are not specified then chi-squared tests are the most powerful if the grouping intervals are chosen in such a way that the probabilities of falling into these intervals are equal or almost equal. This principle is mostly used in this book.

In concluion, we recommend the following method of applying the modified chi-squared test for location and scale families.

If alternatives are not specified and the data are not very roughly rounded then the testing procedure is as follows:

1) Choose equal probabilities $p_i = 1/k$, $i = 1, ..., k$ of falling into grouping intervals. In this case $P_i = i/k$. The number of intervals $k$ is such that $n/k > 5$.

2) Find the ends of the intervals: $a_i = \hat{\mu} + z_i\hat{\sigma}$; where $z_i = G^{-1}(P_i)$, $\hat{\mu}, \hat{\sigma}$ are ML estimators from the initial non-grouped data.

3) Find the numbers $U_i$ of observations falling into the intervals $(a_{i-1}, a_i]$.

4) Compute the statistic

$$X_n^2 = \sum_{i=1}^{k} \frac{U_i^2}{np_i} - n = \frac{k}{n} \sum_{i=1}^{k} U_i^2 - n$$

If $X_n^2 > \chi_\alpha^2(k-1)$ (or, equivalently, $pv_1 = 1 - F_{\chi_{k-1}^2}(X_n^2) < \alpha$) then reject the hypothesis.

5) If $\alpha$ is greater than $pv_1$ (or, equivalently, $X_n^2 < \chi_\alpha^2(k-1)$) then compute the statistic $Q_n$ using formulas [2.35]. $\lambda_3 = 0$ in the case of equal probabilities $p_i = 1/k$ and $g(-x) = g(x)$.

Reject the zero hypothesis if

$$pv_a = 1 - F_{\chi_{k-1}^2}(Y_n^2) < \alpha$$

(or, equivalently, $Y_n^2 = X^2 + Q_n > \chi_\alpha^2(k-1)$).

**Example 2.5.** In 49 locations of the same area of an oil field, quantities $V_i$ of oil were determined. These quantities (in conditional units) are given in the following table.

| 8.7 | 6.6 | 10.0 | 24.3 | 7.9 | 1.3 | 26.2 | 8.3 | 0.9 | 7.1 |
|-----|-----|------|------|-----|-----|------|-----|-----|-----|
| 5.9 | 16.8 | 6.0 | 13.4 | 31.7 | 8.3 | 28.3 | 17.1 | 16.7 | 19.7 |
| 5.2 | 18.9 | 1.0 | 3.5 | 2.7 | 12.0 | 8.3 | 14.8 | 6.3 | 39.3 |
| 4.3 | 19.4 | 6.5 | 7.4 | 3.4 | 7.6 | 8.3 | 1.9 | 10.3 | 3.2 |
| 0.7 | 19.0 | 26.2 | 10.0 | 17.7 | 14.1 | 44.8 | 3.4 | 3.5 | |

Verify the hypotheses: a) the random variables $V_i$ are normally distributed; b) the random variables $V_i$ are lognormally distributed; c) the random variables $V_i^{1/4}$ are normally distributed.

a)

1) Choose $k = 6$, $p_i = 1/6$.

2–3) We have $\bar{X} = 12.018$ and $s = 9.930$. The values of $a_i$ and $U_i$ are given in the following table.

| $i$ | 0 | 1 | 2 | 3 | 4 | 5 | 6 |
|-----|-----|--------|--------|---------|---------|---------|-----|
| $a_i$ | $-\infty$ | 2.4117 | 7.7411 | 12.0180 | 16.2949 | 21.6243 | $+\infty$ |
| $U_i$ | | 5 | 16 | 10 | 3 | 8 | 7 |

4) The statistic $X_n^2 = \frac{6}{49} \sum_{i=1}^{6} U_i^2 - 49 = 12.5918$. Since $k = 6$

$$1 - F_{\chi^2_{k-1}}(X_n^2) = 1 - F_{\chi^2_5}(X_n^2) = 0.0275$$

The $P$-value of the test based on the statistic $Y_n^2$ is less than 0.02752, so the normality hypothesis is rejected and we do not need to compute the statistic $Y_n^2$. Nevertheless, we computed

$Q_n = 12.9907$, $Y_n^2 = 25.5825$ and $pv_a = \mathbf{P}\{\chi_5^2 > 25.5825\} = 0.0001$. The test based on the statistic $Y_n^2$ rejects the hypothesis much more strongly.

b) Let us consider the sample $\ln(V_1), \ldots, \ln(V_{49})$.

1) Choose $k = 6$, $p_i = 1/6$.

2–3) We obtain the following ML estimators of the parameters $\mu$ and $\sigma$: $\bar{X} = 2.1029$ and $s = 0.9675$. The values of $a_i$ and $U_i$ are given in the following table.

| $i$ | 0 | 1 | 2 | 3 | 4 | 5 | 6 |
|---|---|---|---|---|---|---|---|
| $a_i$ | $-\infty$ | 1.1675 | 1.6865 | 2.1030 | 2.5195 | 3.0385 | $+\infty$ |
| $U_i$ | | 7 | 6 | 9 | 9 | 11 | 7 |

4) The statistic $X_n^2 = \frac{6}{49}\sum_{i=1}^{6} U_i^2 - 49 = 2.0612$ and $\mathbf{P}\{\chi_5^2 > 2.0612\} = 0.8406$. We did not find sufficient contradiction to the zero hypothesis.

5) We continue the analysis and compute the value of the statistic $Y_n^2$. Using [2.35], in the case of the normal distribution we obtain $Q_n = 7.3445$ and $Y_n^2 = X_n^2 + Q_n = 9.4057$. The asymptotic $P$-value of the test based on $Y_n^2$ is

$$\mathbf{P}\{\chi_5^2 > 9.4057\} = 0.0939$$

So the data do not contradict the hypothesis if the significance level is greater than 0.0939.

c) Let us consider the sample $V_1^{1/4}, \ldots, V_{49}^{1/4}$.

1) Choose $k = 6$, $p_i = 1/6$.

2–3) We obtain the following ML estimators of the parameters $\mu$ and $\sigma$: $\bar{X} = 1.7394$ and $s = 0.3984$. The values of $a_i$ and $U_i$ are given in the following table.

| $i$ | 0 | 1 | 2 | 3 | 4 | 5 | 6 |
|---|---|---|---|---|---|---|---|
| $a_i$ | $-\infty$ | 1.3539 | 1.5678 | 1.7394 | 1.9110 | 2.1249 | $+\infty$ |
| $U_i$ | | 7 | 8 | 12 | 4 | 11 | 7 |

4) The statistic $X_n^2 = \frac{6}{49} \sum_{i=1}^{6} U_i^2 - 49 = 5.245$ and $\mathbf{P}\{\chi_5^2 > 5.245\} = 0.3867$. We did not find sufficient contradiction to the zero hypothesis.

5) We continue the analysis and compute the value of the statistic $Y_n^2$. Using [2.35], in the case of the normal distribution we obtain: $Q_n = 0.2294$ and $Y_n^2 = X_n^2 + Q_n = 5.4743$. The asymptotic $P$-value of the test based on $Y_n^2$ is

$$\mathbf{P}\{\chi_5^2 > 5.4743\} = 0.3608$$

So the data do not contradict the hypothesis.

**Example 2.6.** Test the hypothesis that $n = 100$ numbers given in the following table are obtained observing normally distributed random variables.

237.34 247.43 251.30 257.64 258.87 261.01 263.05 265.37
265.77 265.95 271.59 273.84 278.85 282.56 283.10 283.18
283.22 285.99 287.81 288.24 291.15 291.86 294.32 295.36
295.47 295.90 296.92 297.63 298.75 300.52 302.95 303.58
304.46 304.55 305.24 305.24 306.25 306.64 306.80 307.31
307.96 308.49 309.70 310.25 310.32 312.18 313.18 313.37
313.61 313.63 315.03 316.35 317.91 318.34 319.42 322.38
324.55 325.02 325.47 326.14 327.82 327.83 337.45 340.73
341.14 342.14 343.50 344.21 346.49 346.72 346.81 348.46
350.19 350.20 351.25 352.23 353.04 353.44 353.79 355.09
355.63 357.48 365.92 366.76 370.46 371.77 373.10 373.92
380.89 381.26 387.94 391.21 402.68 406.04 406.60 409.58
414.93 415.18 418.82 444.66

1) Take $k = 8$ and $p_1 = ... = p_8 = 1/8 = 0.125$.

2) The ML estimators of the parameters $\mu$ and $\sigma$ are $\hat{\mu} = 324.3367$, $\hat{\sigma} = 42.9614$, and

$$P_i = i/8, \quad z_i = \Phi^{-1}(P_i), \quad a_i = \hat{\mu} + z_i \hat{\sigma}$$

The values of $z_i$, $a_i$ and $U_i$ are given in the following table.

| $i$ | 0 | 1 | 2 | 3 | 4 |
|---|---|---|---|---|---|
| $z_i$ | $-\infty$ | -1.1503 | -0.6745 | -0.3186 | 0.0000 |
| $a_i$ | $-\infty$ | 274.919 | 295.360 | 310.650 | 324.337 |
| $U_i$ | – | 12 | 11 | 22 | 11 |

| $i$ | 5 | 6 | 7 | 8 |
|---|---|---|---|---|
| $z_i$ | 0.3186 | 0.6745 | 1.1503 | $+\infty$ |
| $a_i$ | 338.024 | 353.314 | 373.755 | $+\infty$ |
| $U_i$ | 7 | 14 | 10 | 13 |

3) Since

$$X_n^2 = \frac{8}{100} \sum_{i=1}^{6} U_i^2 - 100 = 10.72$$

and

$$\mathbf{P}\{\chi_7^2 > 10.72\} = 0.1513$$

is not sufficiently small, we compute the statistic $Y_n^2$.

4) Using [2.35] as the formulas for the normal distribution, we obtain $Q = 0.3027$, $Y_n^2 = X_n^2 + Q = 11.0227$.

The asymptotic $P$-value of the test based on $Y_n^2$ is $\mathbf{P}\{\chi_7^2 > 11.0227\} = 0.1376$. The hypothesis is not rejected.

## 2.5. Chi-squared test for independence

Suppose that

$$A = \{A_1, \ldots, A_s : A_i \cap A_j = \emptyset, i \neq j = 1, \ldots s, \cup_{i=1}^s A_i = \Omega\}$$

$$B = \{B_1, \ldots, B_r : B_i \cap B_j = \emptyset, i \neq j = 1, \ldots r, \cup_{i=1}^r B_i = \Omega\}$$

are two complete systems of events which can occur during an experiment.

In particular, if the regions of the values of the components $X$ and $Y$ of the random vector $(X, Y)^T$ are divided into subsets $I_i, i = 1, \ldots, s$ and $J_j, j = 1, \ldots, r$, respectively, then

$$A_i = \{X \in I_i\}, \quad B_j = \{Y \in J_j\}$$

For example, the events $A_1, \ldots, A_5$ may mean that a couple has 0, 1, 2, 3 or more than 3 children, respectively ($X$ being the number of children); the events $B_1, \ldots, B_4$ may mean that the monthly incomes of this couple (in euros) are in the intervals

$$[0, 1500], \quad (1500, 3000], \quad (3000, 5000] \quad \text{and} \quad (5000, \infty)$$

respectively ($Y$ being the income per month).

Suppose that $n$ experiments are performed. Denote by $U_{ij}$ the number of experiments where the event $A_i \cap B_j$ occurs. For example, $U_{23}$ may be the number of couples having 2 children and a monthly income in the interval $(3000, 5000]$.

The data are often given in a table with the following form.

| $A_i \backslash B_j$ | $B_1$ | $B_2$ | ... | $B_r$ | $\Sigma$ |
|---|---|---|---|---|---|
| $A_1$ | $U_{11}$ | $U_{12}$ | ... | $U_{1r}$ | $U_{1.}$ |
| $A_2$ | $U_{21}$ | $U_{22}$ | ... | $U_{2r}$ | $U_{2.}$ |
| $\vdots$ | $\vdots$ | $\vdots$ | $\vdots$ | $\vdots$ | $\vdots$ |
| $A_s$ | $U_{s1}$ | $U_{s2}$ | ... | $U_{sr}$ | $U_{s.}$ |
| $\Sigma$ | $U_{.1}$ | $U_{.2}$ | ... | $U_{.r}$ | $n$ |

**Table 2.1.** *The data*

Here

$$U_{i.} = \sum_{j=1}^{r} U_{ij}, \quad i = 1, ..., s; \quad U_{.j} = \sum_{i=1}^{s} U_{ij}, \quad j = 1, ..., r$$

Set

$$\pi_{ij} = \mathbf{P}\{A_i \cap B_j\},$$

$$\pi_{i.} = \sum_{j=1}^{r} \pi_{ij}, \quad \pi_{.j} = \sum_{i=1}^{s} \pi_{ij}, \quad \sum_{i=1}^{s} \pi_{i.} = 1, \quad \sum_{j=1}^{r} \pi_{.j} = 1$$

The statistic

$$\boldsymbol{U} = (U_{11}, ..., U_{1r}, \ U_{21}, ..., U_{2r}, ..., U_{s1}, ..., U_{sr})^T$$

belongs to the following family of multinomial distributions (see [2.2])

$$\mathcal{P} = \{\mathcal{P}_{s \times r}(n, \boldsymbol{\pi}), \quad \boldsymbol{\pi} = (\pi_{11}, ..., \pi_{sr})^T,$$

$$0 < \pi_{ij} < 1, \quad \sum_{i=1}^{s} \sum_{j=1}^{r} \pi_{ij} = 1\}$$

**Independence hypothesis** (of two complete event systems):

$$H_0' : \pi_{ij} = \mathbf{P}\{A_i \cap B_j\} = \mathbf{P}\{A_i\}\mathbf{P}\{B_j\} = \pi_{i.}\pi_{.j}$$

$$i = 1, \ldots, s; \ j = 1, \ldots, r \qquad\qquad [2.37]$$

If the systems $A$ and $B$ are defined by the random variables $X$ and $Y$, as mentioned above, then the following narrower hypothesis may be considered:

**Independence hypothesis** (of two random variables):

$$H_0 : F(x,y) = \mathbf{P}\{X \leq x, Y \leq y\} = F_1(x)F_2(y), \quad \forall x,y \in \mathbf{R}$$
[2.38]

In the considered example the independence hypothesis $H_0$ means that the number of children $X$ and the monthly income $Y$ of a family are independent random variables (these random variables define the random events $A_i$ and $B_j$, respectively).

The alternative to $H_0'$ is

$$H_1' : \pi_{ij} \neq \pi_{i.}\pi_{.j} \quad \text{for some } i,j$$

The ML estimators of the probabilities $\pi_{ij}$ of the polynomial distribution that maximize the likelihood function

$$L(\boldsymbol{\pi}) = \frac{n!}{U_{11}! \cdots U_{sr}!} \prod_{i=1}^{s} \prod_{j=1}^{r} \pi_{ij}^{U_{ij}}$$

are $\hat{\pi}_{ij} = U_{ij}/n$.

Under the hypothesis $H_0'$, the probabilities $\pi_{ij}$ are functions of $s + r - 2$ parameters

$$\boldsymbol{\theta} = (\pi_{1.}, ..., \pi_{s-1.}, \pi_{.1}, ..., \pi_{.r-1})^T$$

i.e.

$$\pi_{ij} = \pi_{ij}(\boldsymbol{\theta}) = \pi_{i.}\pi_{.j}$$

So, under $H_0'$, the likelihood function has the form (see [2.15])

$$\tilde{L}(\boldsymbol{\theta}) = L(\boldsymbol{\pi})|\pi_{ij} = \pi_{i.}\pi_{.j} = \frac{n!}{U_{11}! \cdots U_{sr}!} \prod_{i=1}^{s} \pi_{i.}^{U_{i.}} \prod_{j=1}^{r} \pi_{.j}^{U_{.j}}$$

Its logarithm is

$$\tilde{\ell}(\boldsymbol{\theta}) = C + \sum_{i=1}^{s} U_{i\cdot} \ln \pi_{i\cdot} + \sum_{j=1}^{r} U_{\cdot j} \ln \pi_{\cdot j}$$

so the ML estimators of the parameters $\pi_{i\cdot}$ verify the equations

$$\frac{\partial \tilde{\ell}}{\partial \pi_{i\cdot}} = \frac{U_{i\cdot}}{\pi_{i\cdot}} - \frac{U_{s\cdot}}{\pi_{s\cdot}} = 0, \quad i = 1, 2. \ldots, s-1$$

and, hence, the equations

$$U_{i\cdot} \pi_{s\cdot} = U_{s\cdot} \pi_{i\cdot}, \quad i = 1, 2. \ldots, s$$

Summing with respect to $i = 1, \ldots s$ and using the equations

$$\sum_{i=1}^{s} \pi_{i\cdot} = 1, \quad \sum_{i=1}^{s} U_{i\cdot} = n$$

we obtain the maximum likelihood estimators of the probabilities $\pi_{i\cdot}$

$$\hat{\pi}_{i\cdot} = U_{i\cdot}/n, \quad i = 1, \ldots, s$$

Analogously,

$$\hat{\pi}_{\cdot j} = U_{\cdot j}/n, \quad j = 1, \ldots, r$$

So, under $H_0$ the ML estimators of the probabilities $\pi_{ij} = \pi_{i\cdot} \pi_{\cdot j}$ of the polynomial distribution are

$$\hat{\pi}_{ij} = \pi_{ij}(\hat{\boldsymbol{\theta}}) = \hat{\pi}_{i\cdot} \cdot \hat{\pi}_{\cdot j} = \frac{U_{i\cdot}}{n} \frac{U_{\cdot j}}{n}$$

Using these estimators we obtain the statistic (see [2.17])

$$X_n^2 = \sum_{i=1}^{s} \sum_{j=1}^{r} \frac{(U_{ij} - n\hat{\pi}_{i\cdot}\hat{\pi}_{\cdot j})^2}{n\hat{\pi}_{i\cdot}\hat{\pi}_{\cdot j}} = n \left( \sum_{i=1}^{s} \sum_{j=1}^{r} \frac{U_{ij}^2}{U_{i\cdot}U_{\cdot j}} - 1 \right) \quad [2.39]$$

By Theorem 2.2 the obtained statistic is asymptotically chi-square distributed (as $n \to \infty$) with

$$rs - 1 - (r + s - 2) = (r - 1)(s - 1)$$

degrees of freedom.

**Chi-squared test for independence:** the hypothesis $H_0'$ is rejected with an asymptotic significance level $\alpha$ if

$$X_n^2 > \chi_\alpha^2((r - 1)(s - 1)) \quad\quad\quad [2.40]$$

If the hypothesis $H_0'$ is rejected then the narrower hypothesis $H_0$ is also rejected.

If $s = r = 2$ then the the statistic $X_n^2$ is expressed as

$$X_n^2 = \frac{n(U_{11}U_{22} - U_{12}U_{21})^2}{U_{1\cdot}U_{2\cdot}U_{\cdot 1}U_{\cdot 2}}$$

**Example 2.7. [CRA 46]** The numbers of couples grouped by number of children (factor $A$) and per month income (factor $B$) are presented in Table 2.2. Test the hypothesis of the independence of two factors $A$ and $B$.

| | $[0, 1500]$ | $(1500, 3000]$ | $(3000, 5000]$ | $(5000, \infty)$ | $\Sigma$ |
|---|---|---|---|---|---|
| 0 | 2161 | 3577 | 2184 | 1636 | 9558 |
| 1 | 2755 | 5081 | 2222 | 1052 | 11110 |
| 2 | 936 | 1753 | 640 | 306 | 3635 |
| 3 | 225 | 419 | 96 | 38 | 778 |
| $\geq 4$ | 39 | 98 | 31 | 14 | 182 |
| $\Sigma$ | 6116 | 10928 | 5173 | 3046 | 25263 |

**Table 2.2.** *The data*

From this table we obtain

$$X_n^2 = 25362 \left( \frac{2161^2}{9558 \cdot 6116} + \frac{3577^2}{9558 \cdot 10928} + \dots + \frac{14^2}{182 \cdot 3046} - 1 \right)$$

$$= 568.5$$

Under the independence hypothesis, this statistic is approximately chi-square distributed with $(4-1)(5-1) = 12$ degrees of freedom. So

$$pv_a = \mathbf{P}\{\chi_{12}^2 > 568.5\} < 10^{-16}$$

and the hypothesis is rejected.

## 2.6. Chi-squared test for homogeneity

Suppose that $s$ independent groups of objects are considered. Denote by $n_i$ the number of objects of the $i$-th group of objects. Suppose that

$$\mathcal{B} = \{B_1, \dots, B_r : B_i \cap B_j = \emptyset, i \neq j = 1, \dots r, \cup_{i=1}^r B_i = \Omega\}$$

is a complete group of events. During an experiment, an object is classified by the values of the factor $B$, which takes the values $B_1, \dots, B_r$.

Set

$$\pi_{ij} = \mathbf{P}\{B_j| \text{ the } j\text{-th object belongs to the } i\text{-th group}\} \quad [2.41]$$

For example, suppose that representatives of $s$ different professions perform a psychological test. Any person can receive one of five possible scores: $1, \ldots, 5$. In this case

$$B_j = \{\text{the score is } j\}$$

$j = 1, \ldots, 5$, and $\pi_{ij}$ is the probability that a person of the $i$-th group receives the score $j$. In this case the factor $B$ is the score.

**Homogeneity hypothesis** (with respect to the factor $B$):

$$H_0 : \pi_{1j} = \ldots = \pi_{sj} := \pi_j, \ j = 1, \ldots, r \qquad [2.42]$$

which means that for any fixed $j$ the probability of the event $B_j$ is the same for all groups of objects.

In the case of the considered example, the homogeneity hypothesis means that for any fixed $j$ the probability of representatives of all professions receiving the score $j$ is the same.

**Comment 2.13.** Tests for the homogeneity hypothesis $H_0$ are often used to test the hypothesis of the equality of distributions of random variables.

Suppose that

$$(X_{i1}, \ X_{i2}, \ldots, X_{in_i})^T, \ i = 1, \ldots, s$$

are $s$ independent simple samples. Using the points $-\infty = a_0 < a_1 < \ldots < a_r = +\infty$, divide the abscissas axis into smaller intervals. The event $B_j$ means that an $X$ takes a value from the $j$-th interval $I_j = (a_{j-1}, a_j]$.

Denote by $F_i(x)$ the distribution function of the random variable $X_{ij}$, $i = 1, ..., s$.

**Homogeneity hypothesis** (of independent samples):

$$H_0' : F_1(x) \equiv F_2(x) \equiv \cdots \equiv F_s(x) \qquad [2.43]$$

The hypothesis $H_0'$ is narrower than the hypothesis $H_0$ because the probabilities

$$\pi_{ij} = \mathbf{P}\{X_i \in (a_{j-1}, \ a_j]\} = F_i(a_j) - F_i(a_{j-1}),$$

$$i = 1, ..., s, \ j = 1, ..., r$$

do not depend on $i$ if the hypothesis $H_0'$ holds. So, if the hypothesis $H_0$ is rejected then it is natural to reject the hypothesis $H_0'$. If the hypothesis $H_0$ is accepted then we can say that the grouped data do not contradict the hypothesis $H_0'$.

Denote by $U_{ij}$ the number of objects of the $i$-th group with the value $B_j$ of the factor $B$

$$U_{i1} + \cdots + U_{ir} = n_i$$

i.e. $U_{ij}$ is the number of observations of the $i$-th sample belonging to the interval $I_j$. In the case of the considered example, $U_{ij}$ is the number of representatives of the $i$-th profession which received the score $j$.

The data can be written in a table analogous to Table 2.1.

|  | 1 | 2 | ... | $r$ | $\Sigma$ |
|---|---|---|---|---|---|
| 1 | $U_{11}$ | $U_{12}$ | ... | $U_{1r}$ | $n_1$ |
| 2 | $U_{21}$ | $U_{22}$ | ... | $U_{2r}$ | $n_2$ |
| $\vdots$ | $\vdots$ | $\vdots$ | $\vdots$ | $\vdots$ | $\vdots$ |
| $s$ | $U_{s1}$ | $U_{s2}$ | ... | $U_{sr}$ | $n_s$ |
| $\Sigma$ | $U_{.1}$ | $U_{.2}$ | ... | $U_{.r}$ | $n$ |

**Table 2.3.** *The data*

The random vector $U_i = (U_{i1}, ..., U_{ir})^T$ has the polynomial distribution

$$U_i \sim \mathcal{P}_r(n_i, \pi_i), \quad \pi_i = (\pi_{i1}, ..., \pi_{ir})^T$$

Under the hypothesis $H_0$, the probabilities $\pi_{ij}$ are functions of $r - 1$ parameters $\theta = (\pi_1, ..., \pi_{r-1})^T$. So the likelihood function has the form

$$\tilde{L}(\pi_1, \ldots, \pi_{r-1}) = \prod_{i=1}^{s} \frac{n_i!}{U_{i1}! \cdots U_{ir}!} \pi_1^{U_{i1}} \cdots \pi_r^{U_{ir}} = \prod_{i=1}^{s} \prod_{j=1}^{r} \frac{n_i!}{U_{ij}!} \pi_j^{U_{ij}}$$

$$\pi_r = 1 - \sum_{i=1}^{r-1} \pi_i$$

Its logarithm is

$$\tilde{\ell}(\pi_1, \ldots, \pi_{r-1}) = C + \sum_{i=1}^{s} \sum_{j=1}^{r} U_{ij} \ln \pi_j = C + \sum_{j=1}^{r} U_{\cdot j} \ln \pi_j$$

and the ML estimators of the parameters $\pi_j$ verify the equations

$$\frac{\partial \ell}{\partial \pi_j} = \frac{U_{\cdot j}}{\pi_j} - \frac{U_{\cdot r}}{\pi_r} = 0, \quad j = 1, 2. \ldots, r - 1$$

hence, the equations

$$U_{\cdot j} \pi_r = U_{\cdot r} \pi_j, \quad j = 1, 2. \ldots, r$$

Summing with respect to $j = 1, \ldots r$ and using the equalities

$$\sum_{j=1}^{r} \pi_j = 1, \quad \sum_{i=1}^{s} U_{\cdot j} = n$$

we obtain that the maximum likelihood estimators of the probabilities $\pi_r$. are

$$\hat{\pi}_r = U_{\cdot r}/n$$

Consequently

$$\hat{\pi}_j = U_{.j}/n, \; j = 1, ..., r$$

So, under $H_0$, the ML estimators of the probabilities $\pi_{ij}$ are

$$\hat{\pi}_{ij} = \hat{\pi}_j = \frac{U_{.j}}{n}, \quad j = 1, \ldots, r$$

Using these estimators, we obtain the statistic (see [2.17])

$$X_n^2 = \sum_{i=1}^{s} \sum_{j-1}^{r} \frac{(U_{ij} - n_i\hat{\pi}_j)^2}{n_i\hat{\pi}_j} = n \left( \sum_{i=1}^{s} \sum_{j-1}^{r} \frac{U_{ij}^2}{n_i U_{.j}} - 1 \right) \quad [2.44]$$

Using slightly more general conditions than Theorem 2.2 (one chi-squared type sum of squares is replaced by the sum of several independent chi-squared type sums of squares), asymptotically $(n_i \rightarrow \infty, \; i = 1, ..., s)$ chi-square distributed with

$$s(r - 1) - (r - 1) = (r - 1)(s - 1)$$

degrees of freedom because $r - 1$ parameters were estimated.

**Chi-squared homogeneity test:** the homogeneity hypothesis $H_0$ is rejected with an asymptotic significance level $\alpha$ if

$$X_n^2 > \chi_\alpha^2((r - 1)(s - 1)) \quad [2.45]$$

If the hypothesis $H_0$ is rejected then the narrower hypothesis $H_0'$ is also rejected.

Hypotheses even narrower than $H_0'$ may be considered:

1. Hypothesis

$$H_1' : F_1 = ... = F_s = F_0$$

meaning that all cumulative distribution functions $F_1, ..., F_s$ are equal to a specified cumulative distribution function $F_0$.

In terms of the probabilities $\pi_{ij}$, a wider hypothesis is formulated

$$H_1 : \pi_{1j} = ... = \pi_{sj} = \pi_{j0} = F_0(a_j) - F_0(a_{j-1}), \quad j = 1, ..., r$$

Since there are no unknown parameters, the test statistic is constructed as in section 2.2.

$$X_n^2 = \sum_{i=1}^{s} \sum_{j=1}^{r} \frac{(U_{ij} - n_i \pi_{j0})^2}{n_i \pi_{j0}} \qquad [2.46]$$

If $H_1$ is true and $n_i \rightarrow \infty$, $i = 1, ..., s$ then the interior sums are asymptotically chi-square distributed with $r - 1$ sums of squares, so $X_n^2$, with $s(r - 1)$ degrees of freedom. The hypothesis $H_1$ is rejected if

$$X_n^2 > \chi_\alpha^2(s(r - 1)) \qquad [2.47]$$

The narrower hypothesis $H_1'$ is then also rejected.

2. Hypothesis

$$H_2' : F_1(x) \equiv ... \equiv F_s(x) \equiv F(x, \boldsymbol{\theta})$$

meaning that all cumulative distribution functions $F_1, ..., F_s$ coincide with a cumulative distribution function $F(x; \boldsymbol{\theta})$ of known form depending on a finite-dimensional unknown parameter $\boldsymbol{\theta} = (\theta_1, ..., \theta_l)^T, 0 < l < r.$

In terms of the probabilities $\pi_{ij}$, a wider hypothesis is formulated

$$H_2 : \pi_{1j} = ... = \pi_{sj} = \pi_j(\boldsymbol{\theta}) = F(a_j; \boldsymbol{\theta}) - F(a_{j-1}; \boldsymbol{\theta}), \quad j = 1, ..., r$$

If we estimate the parameter $\theta$ by the polynomial ML or chi-squared minimum method, we obtain the statistic

$$X_n^2 = \sum_{i=1}^{s} \sum_{j=1}^{r} \frac{(U_{ij} - n_i \pi_j(\hat{\boldsymbol{\theta}}))^2}{n_i \pi_j(\hat{\boldsymbol{\theta}})} \qquad [2.48]$$

which, by Theorem 2.2, is asymptotically $(n_i \to \infty)$ chi-square distributed with $s(r-1)-l$ degrees of freedom. The hypothesis $H_2$ is rejected if

$$X_n^2 > \chi_\alpha^2(s(r-1)-l) \qquad [2.49]$$

Then the narrower hypothesis $H_2'$ is also rejected.

**Example 2.8.** [CRA 46] The numbers of children (boys and girls) born in Sweden each month in 1935 [CRA 46] are given in the following table. Test the hypothesis that the probability of a boy birth is constant during the year. Numbers of newborn

| Month | $U_{i1}$ | $U_{i2}$ | $n_i$ | Month | $U_{i1}$ | $U_{i2}$ | $n_i$ |
|---|---|---|---|---|---|---|---|
| 1 | 3743 | 3537 | 7280 | 7 | 3964 | 3621 | 7585 |
| 2 | 3550 | 3407 | 6957 | 8 | 3797 | 3596 | 7393 |
| 3 | 4017 | 3866 | 7883 | 9 | 3712 | 3491 | 7203 |
| 4 | 4173 | 3711 | 7884 | 10 | 3512 | 3391 | 6903 |
| 5 | 4117 | 3775 | 7892 | 11 | 3392 | 3160 | 6552 |
| 6 | 3944 | 3665 | 7609 | 12 | 3751 | 3371 | 7122 |

**Table 2.4.** *The data*

In this table $U_{i1}$ is the number of boys; $U_{i2}$ is the number of girls; $s = 12$, $r = 2$. Under the hypothesis there is only one unknown parameter $p$ – the probability of a boy's birth. Using [2.44] we obtain

$$X_n^2 = \sum_{i=1}^{12} \left( \frac{(U_{i1} - n_i\hat{p})^2}{n_i\hat{p}} + \frac{(U_{i2} - n_i(1 - \hat{p}))^2}{n_i(1 - \hat{p})} \right) = 14.6211$$

$$\hat{p} = \frac{U_{.1}}{n} = \frac{45672}{88263} = 0.51745$$

The number of degrees of freedom is $(s - 1)(r - 1) = 11$. Since

$$pv_a = \mathbf{P}\{\chi_{11}^2 > 14,6211\} = 0.2005$$

we have no basis for rejecting the hypothesis.

## 2.7. Bibliographic notes

The chi-squared test for simple hypotheses was proposed by Pearson [PEA 00]. The chi-squared tests for composite hypotheses were first considered by Fisher and Cramér [FIS 50, PEA 48, CRA 46] and this direction was developed by many statisticians.

The modified chi-squared tests for composite hypotheses were first considered by Nikulin [NIK 73a, NIK 73b, NIK 73c], Rao and Robson [RAO 74], and Dzhaparidze and Nikulin [DZH 74].

This direction was developed by Bolshev and Mirvaliev [BOL 78], Dudley [DUD 79], LeCam *et al.* [LEC 83], Drost [DRO 88, DRO 89], Nikulin and Mirvaliev [NIK 92], Aguiree and Nikulin [AGU 94a, AGU 94b], and many others.

The most recent results were obtained using numerical methods and simulation, where the dependence of the power of tests on different alternatives and grouping methods were considered; see Lemeshko *et al.* [LEM 98b, LEM 98c, LEM 01, LEM 02, LEM 03].

## 2.8. Exercises

**2.1.** Find the first two moments of the statistic $X_n^2$ from [2.5] when the tested hypothesis a) is true; b) is not true.

**2.2.** (continuation of exercise 2.1). Prove that under equal probabilities $\pi_{i0} = 1/k$, the variance of $X_n^2$ is $\mathbf{Var}(X_n^2) =$

$2k^2\{\sum_i \pi_i^2 - (\sum_i \pi_i^2)^2\} + 4(n - 1k^2\{\sum_i \pi_i^3 - \sum_i \pi_i^2\})$, and if $\pi_i = \pi_{i0} = 1/k$ then $\mathbf{Var}(X_n^2) = 2(k-1)$.

**2.3.** Prove that under the assumptions of Theorem 2.1 the statistic

$$\tilde{X}_n^2 = n \sum_{i=1}^{k} (1 - \pi_{i0})[H(U_i/n) - H(\pi_{i0})]^2$$

is asymptotically $(n \to \infty)$ chi-square distributed with $k-1$ degrees of freedom; here $H(x) = \arcsin x$.

**2.4.** Prove that the rank of the matrix $\Sigma = \mathbf{E}_k - \sqrt{\pi}(\sqrt{\pi})^T$ is equal to $k - 1$; here $\mathbf{E}_k$ − the unit $k \times k$ matrix; $\sqrt{\pi} = (\sqrt{\pi_1}, ..., \sqrt{\pi_k})^T$, $0 \le \pi_i \le 1$, $i = 1, ..., k$; $\pi_1 + ... + \pi_k = 1$.

**2.5.** The frequencies of the numbers 0, 1, 2...., 9 among the first 800 digits of the number $\pi$ are respectively 74, 92, 83, 79, 80, 73, 77, 75, 76 and 91. Can these data be interpreted as a realization of a random vector $U \sim \mathcal{P}_{10}(800, \pi)$, $\pi = (1/10, ..., 1/10)$?

**2.6.** The hypothesis on the correctness of random number tables is tested. This hypothesis means that the numbers 0, 1, 2,...,9 appear in the table with equal probabilities $p = 0.1$. Pearson's chi-squared test is used. Find the sample size necessary to reject the hypothesis with a probability not less than 0.95 under the alternative that 5 numbers appear with the probability 0.11, and another 5 with the probability 0.09 (fix the significance level at 0.05).

**2.7.** Peas having yellow and round seeds were interbred with peas having green and wrinkly seeds. The results are given in the table with the probabilities, computed using Mendel's heredity theory.

| Seeds | Frequencies | Probabilities |
|---|---|---|
| Yellow and round | 315 | 9/16 |
| Yellow and wrinkled | 101 | 3/16 |
| Green and round | 108 | 3/16 |
| Green and wrinkled | 32 | 1/16 |
| $\Sigma$ | 556 | 1 |

Do the data contradict Mendel's heredity theory?

**2.8.** Interbreeding two types of maize gives four different types of plant. From Mendel's heredity theory these types should appear with the probabilities 9/16, 3/16, 3/16 and 1/16. Among 1301 plants the following numbers were observed: 773, 231, 238 and 59. Under which significance level does the chi-squared test conclude that the data do not contradict Mendel's model?

**2.9.** On the reading display on the scale of a device the last digit is evaluated by eye [SHE 99]. Sometimes the observers, for no reason, prefer certain digits. The frequencies of the last digits of 200 observations performed by an observer are given in the table.

| Digit | 0 | 1 | 2 | 3 | 4 | 5 | 6 | 7 | 8 | 9 |
|---|---|---|---|---|---|---|---|---|---|---|
| Frequency | 35 | 16 | 15 | 17 | 17 | 19 | 11 | 16 | 30 | 24 |

We see from the table that the digits 0 and 8 have slightly greater frequencies than the other digits. Can we conclude that the observer makes a systematic error?

**2.10.** In 8,000 independent multinomial experiments with three possible issues, the events $A$, $B$ and $C$ appeared 2,014,

5,012 and 974 times. Using the chi-squared test, verify the hypothesis that the probabilities of these events in one experiment are $p_A = 0.5 - 2\alpha$, $p_B = 0.5 + \alpha$ and $p_C = \alpha$, $0 < \alpha < 0.25$.

**2.11.** Among 2,020 families were found 527 with two boys, 476 with two girls and 1,017 with children of both sexes. Test the following hypotheses: a) the number of boys in a family has a binomial distribution; b) the number of boys in a family having two children has a binomial distribution if the probabilities of birth are the same for girls and boys.

**2.12.** Five realizations of a discreet random variable are 47, 46, 49, 53 and 50. Test the hypothesis that this random variable has a Poisson distribution.

**2.13.** [RUT 30] Let us consider the results of the experiment conducted by Rutherford, Geiger and Chadwick [RUT 30] in studies of $\alpha$-decay. The number of scintillations $X_1, ..., X_n$ were registered on a zinc sulfide screen as a result of penetration into the screen by the $\alpha$-particles emitted from a radium source in $n = 2608$ mutually exclusive time intervals, each of duration 7.5 s. Let $n_j$ represent the frequency of the event $\{X = j\}$, i.e. $n_j$ is the number of values of $X_i$ equal to $j$. The frequencies $n_j$ from Rutherford's data are given in the following table.

| $j$ | 0 | 1 | 2 | 3 | 4 | 5 | 6 | 7 | 8 | 9 | 10 | 11 | 12 |
|---|---|---|---|---|---|---|---|---|---|---|---|---|---|
| $n_j$ | 57 | 203 | 383 | 525 | 532 | 408 | 273 | 139 | 45 | 27 | 10 | 4 | 2 |

Do the data contradict the hypothesis that the number of scintillations has a Poisson distribution?

**2.14.** Using control devices, the distance $r$ (in microns) from the center of gravity to the axis of its external cylinder was measured. The results are given in the table ($r_i$ – values, $n_i$ – frequencies).

| $r_i$ | $n_i$ | $r_i$ | $n_i$ |
|---|---|---|---|
| $0 - 16$ | 40 | $80 - 96$ | 45 |
| $16 - 32$ | 129 | $96 - 112$ | 19 |
| $32 - 48$ | 140 | $112 - 128$ | 8 |
| $48 - 64$ | 126 | $128 - 144$ | 3 |
| $64 - 80$ | 91 | $144 - 160$ | 1 |

Apply the chi-squared test to verify the hypothesis that the distance has a Relay distribution.

**2.15.** The data on the failure times of 200 electric bulbs are given in the table ($(a_{i-1}, a_i]$ is the failure time interval, $n_i$ is the frequency).

| $(a_{i-1}, a_i]$ | $n_i$ | $(a_{i-1}, a_i]$ | $n_i$ |
|---|---|---|---|
| $0 - 300$ | 53 | $1800 - 2100$ | 9 |
| $300 - 600$ | 41 | $2100 - 2400$ | 7 |
| $600 - 900$ | 30 | $2400 - 2700$ | 5 |
| $900 - 1200$ | 22 | $2700 - 3000$ | 3 |
| $1200 - 1500$ | 16 | $3000 - 3300$ | 2 |
| $1500 - 1800$ | 12 | $3300 - 3600$ | 0 |

Apply the chi-squared test to verify the hypothesis that the failure time of a bulb has an exponential distribution.

**2.16.** The realization of a sample of size $n = 100$ is given in the table.

| 338 | 336 | 312 | 322 | 381 | 302 | 296 | 360 | 342 | 334 |
|-----|-----|-----|-----|-----|-----|-----|-----|-----|-----|
| 348 | 304 | 323 | 310 | 368 | 341 | 298 | 312 | 322 | 350 |
| 304 | 302 | 336 | 334 | 304 | 292 | 324 | 331 | 324 | 334 |
| 314 | 338 | 324 | 292 | 298 | 342 | 338 | 331 | 325 | 324 |
| 326 | 314 | 312 | 362 | 368 | 321 | 352 | 304 | 302 | 332 |
| 314 | 304 | 312 | 381 | 290 | 322 | 326 | 316 | 328 | 340 |
| 324 | 320 | 364 | 304 | 340 | 290 | 318 | 332 | 354 | 324 |
| 304 | 321 | 356 | 366 | 328 | 332 | 304 | 282 | 330 | 314 |
| 342 | 322 | 362 | 298 | 316 | 298 | 332 | 342 | 316 | 326 |
| 308 | 321 | 302 | 304 | 322 | 296 | 322 | 338 | 324 | 323 |

Apply the modified chi-squared test (number of grouping intervals $k = 8$) to verify the hypothesis that a random variable having a normal distribution was observed.

**2.17.** The concentrations of an element in the residue after a chemical reaction are given in the table.

| 10 | 51 | 8 | 47 | 8 | 5 | 56 | 12 | 4 | 5 | 4 | 4 | 7 | 6 | 9 |
|----|----|----|----|----|----|----|----|----|----|----|----|----|----|----|
| 30 | 25 | 12 | 3 | 22 | 5 | 15 | 4 | 4 | 29 | 15 | 4 | 2 | 18 | 41 |
| 3 | 5 | 54 | 110 | 24 | 16 | 2 | 37 | 20 | 2 | 6 | 7 | 16 | 2 | 14 |
| 68 | 10 | 16 | 11 | 78 | 6 | 17 | 7 | 11 | 21 | 15 | 24 | 6 | 32 | 8 |
| 11 | 4 | 14 | 45 | 17 | 10 | 15 | 20 | 4 | 65 | 10 | 3 | 5 | 11 | 13 |
| 35 | 11 | 34 | 3 | 4 | 12 | 7 | 6 | 62 | 13 | 36 | 26 | 6 | 11 | 6 |
| 13 | 1 | 4 | 36 | 18 | 10 | 37 | 28 | 4 | 12 | 31 | 14 | 3 | 11 | 6 |
| 4 | 10 | 38 | 6 | 11 | 24 | 9 | 4 | 5 | 8 | 135 | 22 | 6 | 18 | 49 |
| 17 | 9 | 32 | 27 | 2 | 12 | 8 | 93 | 3 | 9 | 10 | 3 | 14 | 33 | 72 |
| 14 | 4 | 9 | 10 | 19 | 2 | 5 | 21 | 8 | 25 | 30 | 20 | 12 | 19 | 16 |

Apply the modified chi-squared test (number of grouping intervals $k = 10$) to verify the hypothesis that the concentration has a lognormal distribution.

**2.18.** In cells influenced by $X$-rays some chromosomes mutate. The data gathered in several series of independent experiments ($i$ is the number of chromosome mutations, $n_{ik}$ is the number of cells with $i$ mutations in the $k$-th experiment) are given in the table.

| $i$ | 0 | 1 | 2 | $\geq 3$ | $\sum n_{ik}$ |
|---|---|---|---|---|---|
| $n_{i1}$ | 280 | 75 | 12 | 1 | 368 |
| $n_{i2}$ | 593 | 143 | 20 | 3 | 759 |
| $n_{i3}$ | 639 | 141 | 13 | 0 | 793 |
| $n_{i4}$ | 359 | 109 | 13 | 1 | 482 |

Test the hypothesis that all four samples are obtained by observing random variables with a) Poisson distribution; b) the same Poisson distribution.

**2.19.** Test the hypothesis that $n = 100$ numbers given in the table are realizations of a random variable having a normal distribution.

24 41 30 37 25 32 28 35 28 51 36 26 43 25 27 39 21 45 39 25 29
43 66 25 24 56 29 31 41 41 36 57 36 48 25 36 48 24 48 22 40 7
31 24 32 53 33 46 22 33 25 37 34 32 41 36 19 32 25 19 19 37 20
21 48 44 35 19 44 34 29 48 38 43 48 35 42 37 35 36 58 45 34 40
37 21 41 11 41 27 50 24 37 39 33 45 39 43 21 34

Modify the computation taking into account rounding of the data. Choose $k = 8$.

**2.20.** In two independent samples, each of size 500, the times shown by clocks displayed in windows of various shops were registered [CRA 46]. The data grouped into 12 intervals (0 means the interval from 0 h to 1 h; 1 from 1 h to 2 h, and so on) are given in the table.

| | 1 | 2 | 3 | 4 | 5 | 6 | 7 | 8 | 9 | 10 | 11 | 12 | Σ |
|---|---|---|---|---|---|---|---|---|---|---|---|---|---|
| 1 | 41 | 34 | 54 | 39 | 49 | 45 | 41 | 33 | 37 | 41 | 47 | 39 | 500 |
| 2 | 36 | 47 | 41 | 47 | 49 | 45 | 32 | 37 | 40 | 41 | 37 | 48 | 500 |

Apply the chi-squared test to verify the hypothesis that in both samples the times shown by the clocks fall into any interval with the same probability.

**2.21.** In a group of 300 students, the numbers of students who got the comments on an exam "not sufficient", "sufficient", "good" and "very good" were 33, 43, 80 and 144, respectively; in a second group the respective numbers of students were 39, 35, 72 and 154. Are both groups equally prepared for the exam?

**2.22.** Launching a rocket 87 times, the following data on the distance $X$(m) and deviation $Y$(minutes of angle) were collected. These data are given in the following table.

| $x_i/y_j$ | (-250,-50) | (-50.50) | (50.250) | Σ |
|---|---|---|---|---|
| 0 − 1200 | 5 | 9 | 7 | 21 |
| 1200 − 1800 | 7 | 5 | 9 | 21 |
| 1800 − 2700 | 8 | 21 | 16 | 45 |
| Σ | 20 | 35 | 32 | 87 |

Are the random variables $X$ and $Y$ independent?

**2.23.** Investigating the granular composition of quartz in Lithuanian and Saharan sand samples, the data on the lengths of the maximal axis of the quartz grains were used. Using the granular composition of the samples, conclusions on the geologicical sand formation conditions were drawn. The data are grouped in intervals of equal length ($X_i$ is the center of the $i$-th interval).

| $X_i$ | 9 | 13 | 17 | 21 | 25 | 29 | 33 | 37 | 41 | 45 | 49 | $\Sigma$ |
|---|---|---|---|---|---|---|---|---|---|---|---|---|
| Lithuanian sand | 4 | 12 | 35 | 61 | 52 | 23 | 7 | 4 | 2 | 1 | 0 | 201 |
| Sahara sand | 0 | 6 | 10 | 12 | 13 | 12 | 15 | 12 | 11 | 7 | 4 | 102 |

Apply a chi-squared test to verify the hypothesis that the length of the maximum axis has the same distribution in Lithuanian and Saharan sand.

## 2.9. Answers

**2.1.** a) $\mathbf{E}(X_n^2) = k - 1$, $\mathbf{Var}(X_n^2) = 2(k - 1 + [2(k - 1) - k^2 + \sum_i(1/\pi_{i0})])/n = 2(k - 1) + O(1/n)$. b) $\mathbf{E}(X_n^2) = k - 1 + n\sum_i((\pi_i - \pi_{i0})^2/\pi_{i0}) + \sum_i((\pi_i - \pi_{i0})(1 - \pi_i)/\pi_{i0})$; $\mathbf{Var}(X_n^2) = (2(n - 1)/n)\{2(n - 2)\sum_i(\pi_i^3/\pi_{i0}^2) - (2n - 3)(\sum_i(\pi_i^2/\pi_{i0}))^2 - 2\sum_i(\pi_i^2/\pi_{i0})\sum_i(\pi_i/\pi_{i0}) + 3\sum_i(\pi_i^2/\pi_{i0}^2)\} - [(\sum_i(\pi_i/\pi_{i0}))^2 - \sum_i(\pi_i/\pi_{i0}^2)]/n$. **Note.** $\mathbf{Var}(X_n^2) = \mathbf{E}(\sum_i(U_i^2/(n\pi_{i0})))^2 - (\sum_i(\mathbf{E}U_i^2/(n\pi_{i0})))^2 = \sum_i(\mathbf{E}U_i^4/(n^2\pi_{i0}^2)) + \sum_{i \neq j}(\mathbf{E}(U_i^2 U_j^2)/(n^2\pi_{i0}\pi_{j0})) - (\sum_i(\mathbf{E}U_i^2/(n\pi_{i0}))^2$. Find the moments $\mathbf{E}(U_i^2)$, $\mathbf{E}(U_i^4)$, $\mathbf{E}(U_i^2 U_j^2)$ (using, for example, their expressions in terms of factorial moments)) and simplify.

**2.3. Note.** Prove that the random vector

$$\sqrt{n}(\sqrt{1 - \pi_{10}}(H(U_1/n) - H(\pi_{10})), ..., \sqrt{1 - \pi_{k0}}(H(U_k/n) - H(\pi_{k0})))^T$$

is asymptotically $k$-dimensional normal with a zero mean vector and the same covariance matrix $\Sigma$, as in Theorem 2.1.

**2.5.** The value of the statistic $X_n^2$ is 5.125 and $pv_a = \mathbf{P}\{\chi_9^2 > 5.125\} = 0.8233$; the data do not contradict the hypothesis.

**2.6.** $n \geq 881$.

**2.7.** The value of the statistic $X_n^2$ is 0.47 and $pv_a = \mathbf{P}\{\chi_3^2 > 0.47\} = 0.9254$; the data do not contradict the hypothesis.

**2.8.** The value of the statistic $X_n^2$ is 9.2714 and $pv_a = \mathbf{P}\{\chi_3^2 > 9.2714\} = 0.0259$; the hypothesis is rejected if the significance level is greater than 0.0259.

**2.9.** Testing the hypothesis $H : \pi_i = 1/10, i = 1, ..., 10$, the value of the statistic $X_n^2$ is 24.9 and $pv_a = \mathbf{P}\{\chi_9^2 > 24.9\} = 0.0031$. The hypothesis is rejected. Let us verify the hypothesis that the probability of simultaneous appearance of the numbers 0 and 8 is 0.2. Under this hypothesis $S = U_1 + U_8 \sim B(n, 0.2)$. The value of the sum $S$ is 65. So $pv = \mathbf{P}\{S \geq 65\} = 0.00002$. The conclusion: the observer makes a systemic error.

**2.10.** The estimator of the parameter $\alpha$ is $\hat{\alpha} = 0.1235$. The value of the statistic $X_n^2(\hat{\alpha})$ is 0.3634 and $pv_a = \mathbf{P}\{\chi_1^2 > 0.3634\} = 0.5466$. The data do not contradict the hypothesis.

**2.11.** The estimator of the probability of the birth of a boy is $\hat{p} = 0.5126$. The value of the statistic $X_n^2(\hat{p})$ is 0.1159 and $pv_a = \mathbf{P}\{\chi_1^2 > 0.1159\} = 0.7335$. The data do not contradict the hypothesis. If the probability of the birth of a boy is not estimated and it is supposed that is equal to 1/2, then the value of the statistic $X_n^2$ is 2.6723 and $pv_a = \mathbf{P}\{\chi_2^2 > 2.6723\} = 0.2629$. Again, the data do not contradict the hypothesis.

**2.12.** The statistic $X_n^2$ is asymptotically chi-square distributed with four degrees of freedom and takes the value 0.6122, $pv_a = \mathbf{P}\{\chi_4^2 > 0.6122\} = 0.9617$. The data do not contradict the hypothesis. **Note.** Use the fact: the conditional distribution of a simple sample $(X_1, ..., X_n)^T$, given the sum $S = X_1 + ... + X_n$, is multinomial $\mathcal{P}_n(S, \pi_0), \pi_0 = (1/n, ..., 1/n)^T$ (see section 5.5.4).

**2.13.** The estimator of the parameter $\lambda$ is $\hat{\lambda} = 3.8666$. The value of the statistic $X_n^2(\hat{\lambda})$ is 13.0146, $pv_a = \mathbf{P}\{\chi_{10}^2 > 13.0146\} = 0.2229$. The data do not contradict the hypothesis. Computing the value of the statistic, the two last intervals were unified.

**2.14.** The value of the ML estimator of the parameter $\sigma^2$ is $\hat{\sigma}^2 = 1581.65$, the value of the test statistic is $X_n^2(\hat{\sigma}^2) = 2.6931$ and the $P$-value is $pv_a = \mathbf{P}\{\chi_7^2 > 2.6931\} = 0.9119$. The data do not contradict the hypothesis. Computing the value of the statistic, the two last intervals were unified.

**2.15.** The value of the ML estimator of the parameter $\theta$ is $\hat{\theta} = 878.4$. The value of the test statistic is $X_n^2(\hat{\theta}) = 4.0477$ and the $P$-value is $pv_a = \mathbf{P}\{\chi_8^2 > 4.0477\} = 0.8528$. The data do not contradict the hypothesis. Computing the value of the statistic, the three last intervals were unified.

**2.16.** The values of the ML estimators of the parameter $\mu$ and $\sigma$ are $\hat{\mu} = \bar{X} = 324.57, \hat{\sigma} = 20.8342$. Fix $k = 8$ intervals. Then $X_n^2 = 8.0, Q_n = 4.1339, Y_n^2 = 12.1339$ and $pv_a = \mathbf{P}\{\chi_7^2 > 12.1339\} = 0.0962$. The hypothesis is rejected if the significance level is greater than 0.0962.

**2.17.** Using logarithms of $Y_i = \ln(X_i)$, we obtain the values of the ML estimators $\hat{\mu} = \bar{Y} = 2.4589, \hat{\sigma} = 0.9529$. Fix $k = 10$ intervals. Then $X_n^2 = 4.1333, Q_n = 1.1968, Y_n^2 = 5.3301$ and $pv_a = \mathbf{P}\{\chi_9^2 > 5.3301\} = 0.8050$. The data do not contradict the hypothesis.

**2.18.** a) From each sample estimate the parameter $\lambda$, compute the values of the statistics $X_{n_i}^2(\hat{\lambda})$ and add them. Under the hypothesis this value is a realization of the statistic having an asymptotically chi-squared distribution with four degrees of freedom (we unify the last two intervals in each sample). This value is 2.5659, $pv_a = \mathbf{P}\{\chi_4^2 > 2.5659\} = 0.6329$, so the data do not contradict the hypothesis. b) The estimator of the parameter $\lambda$ takes the value $\hat{\lambda} = 0.2494$. Compute the values of the statistics $X^2_{n_i}(\hat{\lambda}_i)$ and add them. We have 10.2317 and $pv_a = \mathbf{P}\{\chi^2_7 > 10.2317\} = 0.1758$. The data do not contradict the hypothesis.

**2.19.** If we do not take rounding of the data into account then we have $X^2_n = 4.160$, $Q_n = 0.172$, $Y^2_n = 4.332$ and $pv_a = \mathbf{P}\{\chi^2_7 > 4.332\} = 0.741$. Taking the rounding of the data into account, we have $X^{2\prime}_n = 3.731$, $Q'_n = 0.952$, $Y^{2\prime}_n = 4.683$ and $pv'_a = \mathbf{P}\{\chi^2_7 > 4.683\} = 0.699$. The data do not contradict the hypothesis. The $P$-values $pv$ and $pv'$ are considerably different. **Note.** The data are rounded to integers so we shift the points $a'_i$ to the nearest points of the form $m \pm 0.5$ ($m$ is an integer). The points $a'_i$ are used for computing $z'_i$ and $p'_i$. The obtained probabilities $p'_i$ are not equal.

**2.20.** The value of the statistic [2.44] is 8.51 and $pv_a = \mathbf{P}\{\chi^2_{11} > 8.51\} = 0.6670$. The data do not contradict the hypothesis.

**2.21.** The value of the statistic [2.44] is 2.0771 and $pv_a = \mathbf{P}\{\chi^2_3 > 2.0771\} = 0.5566$. The data do not contradict the hypothesis.

**2.22.** The value of the statistic [2.39] is 3.719 and $pv_a = \mathbf{P}\{\chi^2_4 > 3.719\} = 0.4454$. The data do not contradict the hypothesis.

**2.23.** The value of the statistic [2.44] is 75.035 (the first two and the last three intervals are unified), $pv_a = \mathbf{P}\{\chi^2_7 > 75.035\} < 10^{-12}$. The hypothesis is rejected.

# Chapter 3

# Goodness-of-fit Tests Based on Empirical Processes

Suppose that $\mathbf{X} = (X_1, ..., X_n)^T$ is a simple sample of the random variable $X$ with the cdf $F$ belonging to the set $\mathcal{F}$ of absolutely continuous distribution functions. Let us consider the **simple hypothesis**

$$H_0 : F(x) \equiv F_0(x) \qquad [3.1]$$

where $F_0$ is a specified cdf.

## 3.1. Test statistics based on the empirical process

**Idea of tests based on empirical processes.** Suppose that

$$\hat{F}_n(x) = \frac{1}{n} \sum_{i=1}^{n} \mathbf{1}_{(-\infty, x]}(X_i)$$

is the empirical distribution function.

By the Glivenko–Cantelli theorem

$$\sup_{x \in \mathbf{R}} |\hat{F}_n(x) - F_0(x)| \overset{a.s.}{\to} 0 \quad \text{as} \quad n \to \infty$$

under the hypothesis $H_0$.

So tests for the hypothesis $H_0$ based on functionals of the empirical process $\mathcal{E}_n = \sqrt{n}(\hat{F}_n - F_0)$ may be constructed if the probability distribution of these functionals is parameter free.

**Test statistics.** Let us consider the following functionals

$$D_n = \sup_{x \in \mathbf{R}} |\hat{F}_n(x) - F_0(x)| \quad \textit{(Kolmogorov–Smirnov statistic)}$$

[3.2]

$$C_n = \int_{-\infty}^{\infty} (\hat{F}_n(x) - F_0(x))^2 dF_0(x) \quad \textit{(Cramér-von-Mises statistic)}$$

[3.3]

$$A_n = \int_{-\infty}^{\infty} \frac{(\hat{F}_n(x) - F_0(x))^2}{F_0(x)(1 - F_0(x))} dF_0(x) \quad \textit{(Andersen-Darling statistic)}$$

[3.4]

or, generalizing the last two

$$\omega_n^2 = \omega_n^2(\psi) = \int_{-\infty}^{\infty} (\hat{F}_n(x) - F_0(x))^2 \psi(F_0(x)) dF_0(x) \quad \textit{($\omega^2$ statistic)}$$

[3.5]

where $\psi$ is a non-negative function defined on the interval $(0, 1)$.

**Theorem 3.1.** *Suppose that* $\mathbf{X} = (X_1, ..., X_n)^T$ *is a simple sample of an absolutely continuous random variable $X$ with the cdf $F_0$. Then the distribution of any of the statistics [3.2]– [3.5] does not depend on $F_0$ and depends only on the sample size $n$.*

**Proof.** If $X$ is an absolutely continuous random variable then the random variable $Y = F_0(X)$ is uniformly distributed in the interval $[0, 1]$. Set $Y_i = F_0(X_i) \sim U(0, 1)$. The distribution of the empirical distribution function

$$\hat{G}_n(y) = \frac{1}{n} \sum_{i=1}^{n} \mathbf{1}_{(-\infty, y]}(Y_i)$$

does not depend on $F_0$.

Set $u(x) = \sup\{v : F_0(v) = F_0(x)\}$. Since the function $F_0$ does not decrease, we have

$$\mathbf{P}\{X_i \leq x\} = \mathbf{P}\{X_i \leq u(x)\} = \mathbf{P}\{F_0(X_i) \leq F_0(x)\} = \mathbf{P}\{Y_i \leq F_0(x)\}$$

so $\hat{F}_n(x) \overset{d}{=} \hat{G}_n(F_0(x))$; here $\overset{d}{=}$ means equality in distribution.

The function $F_0$ is continuous. So if the points $x$ fill the interval $[-\infty, \infty]$ then the point $y = F_0(x)$ fills the interval $[0, 1]$, hence

$$D_n = \sup_{x \in \mathbf{R}} |\hat{F}_n(x) - F_0(x)| \overset{d}{=} \sup_{x \in \mathbf{R}} |\hat{G}_n(F_0(x)) - F_0(x)| =$$

$$\sup_{y \in [0, 1]} |\hat{G}_n(y) - y|$$

$$\omega_n^2 = \int_{-\infty}^{\infty} (\hat{G}_n(F_0(x)) - F_0(x))^2 \psi(F_0(x)) dF_0(x) =$$

$$\int_0^1 (\hat{G}_n(y) - y)^2 \psi(y) dy$$

This shows that the distribution of $D_n$ and $\omega^2$ does not depend on $F_0$ but depends on the sample size $n$.

$\triangle$

Theorem 3.1. implies that the hypothesis $H_0$ should be rejected if the statistics [3.2]–[3.5] take large values.

If $n$ is large then the probability distribution of the test statistics is complicated and asymptotic distributions of these statistics are needed.

Since all statistics are functionals of the empirical process

$$\mathcal{E}_n(x) = \sqrt{n}(\hat{F}_n(x) - F_0(x)), \quad x \in \mathbf{R} \qquad [3.6]$$

the weak invariance property of the empirical process (see Appendix B, Theorem B.1) is useful in seeking the limit probability distribution of these statistics. By Theorem B.1, the Kolmogorov–Smirnov, Cramér–von-Mises and Andersen–Darling statistics have the following limits

$$\sqrt{n}D_n \xrightarrow{d} \sup_{0 \le t \le 1} |B(t)|, \quad nC_n \xrightarrow{d} \int_0^1 B^2(t)dt$$

$$nA_n \xrightarrow{d} \int_0^1 \frac{B^2(t)}{t(1-t)} dt \qquad [3.7]$$

where $B$ is the Brownian bridge (see Appendix B, sections B.2–B.6).

So, to find the limit distribution, the given functionals of the Brownian bridge should be investigated.

**Discrete distributions.** Suppose that $X$ is a discrete random variable taking the values $a_1, ..., a_k$ with the probabilities $p_1, ..., p_k, \sum_i p_i = 1$. Its cdf is piecewise-constant with jumps $p_i$ at the points $a_i, i = 1, ..., k$.

Suppose that $(X_1, ..., X_n)^T$ is a simple sample of size $n$ and the value $a_i$ repeats $U_i$ times, $\sum_i U_i = n$. Then the empirical distribution function $\hat{F}_n(x)$ has jumps $U_i/n$ at the points $a_i, i = 1, ..., k$.

Let us consider hypothesis [3.2], where $F_0(x)$ is a known discrete cdf of a random variable taking values $a_1, ..., a_k$ with known probabilities $p_{i0}, \sum_i p_{i0} = 1$.

Define discrete analogs of the Kolmogorov–Smirnov and $\omega^2$-type statistics

$$\tilde{D}_n = \max_{1 \le i \le k} |\hat{F}_n(x_i) - F_0(x_i)| = \max_{1 \le i \le k} |\hat{G}_n(t_i) - t_i|$$

$$\tilde{\omega}_n(\psi) = \sum_{i=1}^{k} (\hat{F}_n(x_i) - F_0(x_i))^2 \psi(F_0(x_i)) p_{0i} =$$

$$\sum_{i=1}^{k} (\hat{G}_n(t_i) - t_i)^2 \psi(t_i) p_{0i}$$

where $\hat{G}_n(t)$ is the empirical distribution function with jumps $U_i/n$ at the points $t_i = p_{10} + ... + p_{i0}, i = 1, ..., k$.

The properties of the empirical process are (Appendix B, Theorem B.1)

$$\sqrt{n}\tilde{D}_n \xrightarrow{d} \max_{1 \le i \le k} |B(t_i)|, \quad n\tilde{\omega}_n(\psi) \xrightarrow{d} \sum_{i=1}^{k} B^2(t_i)\psi(t_i) p_{i0}$$

The critical values of the statistics $\tilde{D}_n, \tilde{\omega}_n(\psi)$ under the zero hypothesis can be found by simulation using the fact that the random variables $(U_1, ..., U_k)^T \sim \mathcal{P}(n, \boldsymbol{p}_0), \boldsymbol{p}_0 = (p_{10}, ..., p_{k0})^T$. The asymptotic critical values can be found using the fact that the random vector $(B(t_1), ..., B(t_k))^T$ has a $k$-dimensional normal distribution with zero mean and the covariances $\sigma_{ij} = t_i(1 - t_j), t_i \le t_j$ (Appendix B, section B.2.4).

If the number of values $k$ is large, and all probabilities $p_{i0}$ are small, then the tests are applied as in the continuous case. Note that even in the continuous case the data are rounded with a certain precision, so in fact discrete random variables taking a large number of values with small probabilities are observed (see also exercises 3.6–3.10).

## 3.2. Kolmogorov–Smirnov test

The Kolmogorov–Smirnov test for the hypothesis $H_0$ : $F(x) \equiv F_0(x)$ is based on the statistic

$$D_n = \sup_{x \in \mathbf{R}} |\hat{F}_n(x) - F_0(x)| \qquad [3.8]$$

**Two-sided alternative**

$$\bar{H} : \sup_{x \in \mathbf{R}} |F(x) - F_0(x)| > 0 \qquad [3.9]$$

So deviation from the hypothesis is measured using uniform metrics.

In Theorem 3.1 it was shown that the distribution of the statistic $D_n$ does not depend on $F_0$.

**Computing the Kolmogorov–Smirnov statistic.** The following Theorem is useful for computing the statistic $D_n$.

**Theorem 3.2.** *Suppose that $X_{(1)} \leq \cdots \leq X_{(n)}$ are order statistics. Then the equality*

$$D_n = max(D_n^+, D_n^-\} \qquad [3.10]$$

*holds; here*

$$D_n^+ = \max_{1 \leq i \leq n} [\hat{F}_n(X_{(i)}) - F_0(X_{(i)})]$$

$$D_n^- = \max_{1 \leq i \leq n} [F_0(X_{(i)}) - \hat{F}_n(X_{(i-1)})]$$

*If $X_{(1)} < \cdots < X_{(n)}$ then*

$$D_n^+ = \max_{1 \leq i \leq n} \left( \frac{i}{n} - F_0(X_{(i)}) \right), \quad D_n^- = \max_{1 \leq i \leq n} \left( F_0(X_{(i)}) - \frac{i-1}{n} \right)$$

$$[3.11]$$

**Proof.** If $X_{(i-1)} < X_{(i)}$ then using the fact that $F_0(x)$ is not decreasing in the interval $(X_{(i-1)}, X_{(i)}]$, $\hat{F}_n(x) = \hat{F}_n(X_{(i-1)})$ for all $x \in (X_{(i-1)}, X_{(i)})$, we have

$$\sup_{x \in [X_{(i-1)}, X_{(i)}]} [\hat{F}_n(x) - F_0(x)] = max[\hat{F}_n(X_{(i-1)}) - F_0(X_{(i-1)}),$$

$$\hat{F}_n(X_{(i)}) - F_0(X_{(i)})]$$

$$\sup_{x \in [X_{(i-1)}, X_{(i)}]} [F_0(x) - \hat{F}_n(x)] = F_0(X_{(i)}) - \hat{F}_n(X_{(i-1)})$$

The last two formulas are also true if $X_{(i-1)} = X_{(i)}$. Hence

$$\sup_{x \in \mathbf{R}}[\hat{F}_n(x) - F_0(x)] = \max_{1 \le i \le n} [\hat{F}_n(X_{(i)}) - F_0(X_{(i)})] = D_n^+$$

$$\sup_{x \in \mathbf{R}}[F_0(x) - \hat{F}_n(x)] = \max_{1 \le i \le n} [F_0(X_{(i)}) - \hat{F}_n(X_{(i-1)})] = D_n^-$$

$$D_n = max\{ \sup_{\hat{F}_n(x) - F_0(x) \ge 0} |\hat{F}_n(x) - F_0(x)|,$$

$$\sup_{\hat{F}_n(x) - F_0(x) < 0} |\hat{F}_n(x) - F_0(x)|\} =$$

$$max\{\sup_{x \in \mathbf{R}}[\hat{F}_n(x) - F_0(x)], \sup_{x \in \mathbf{R}}[F_0(x) - \hat{F}_n(x)]\} = max(D_n^+, D_n^-)$$

If $X_{(1)} < \cdots < X_{(n)}$ then $\hat{F}_n(X_{(i)}) = i/n$, hence [3.11] holds.

$\triangle$

**Kolmogorov–Smirnov test:** the hypothesis $H_0$ is rejected with a significance level $\alpha$ if $D_n > D_\alpha(n)$; here $D_\alpha(n)$ is the $\alpha$-critical value of the statistic $D_n$.

For small values of $n$, most statistical software computes the values of the cdf $F_{D_n}$ of the Kolmogorov–Smirnov statistic, so the $P$-value is

$$pv = 1 - F_{D_n}(d) \qquad [3.12]$$

where $d$ is the observed value of $D_n$. If $n$ is large then an asymptotic distribution of the test statistic is used.

**Asymptotic distribution of the statistic** $D_n$. Let us find the asymptotic distribution of the statistic $\sqrt{n}D_n$ using the relation [3.7].

**Theorem 3.3.** *Suppose that* $X_1, \ldots, X_n$ *is a simple sample of a random variable* $X$ *with an absolutely continuous cumulative distribution function* $F_0(x)$. *If* $n \to \infty$ *then for all* $x \in \mathbf{R}$

$$\mathbf{P}\{\sqrt{n}D_n \leq x\} \to K(x) = \sum_{k=-\infty}^{\infty} (-1)^k e^{-2k^2 x^2} =$$

$$1 + 2\sum_{k=1}^{\infty}(-1)^k e^{-2k^2 x^2} \qquad [3.13]$$

**Proof.** The result of the theorem is implied by [3.7] and by property 4 (Appendix B, section B.6) of the Brownian bridge: for all $x > 0$

$$\mathbf{P}\{\sup_{0 \leq t \leq 1} |B_t| \geq x\} = 2\sum_{n=1}^{\infty}(-1)^{n-1} e^{-2n^2 x^2} \qquad [3.14]$$

$\triangle$

Most statistical software computes the values of the cdf $K$. Theorem 3.3 implies that the asymptotic $P$-value of the Kolmogorov–Smirnov test is

$$pv_a = 1 - K(\sqrt{n}D_n)$$

**Simple approximation of critical values.** If $n > 100$ then $D_\alpha(n)$ can be approximated by (see [BOL 83])

$$D_\alpha(n) \approx \sqrt{\frac{1}{2n}\left(y - \frac{2y^2 - 4y - 1}{18n}\right)} - \frac{1}{6n} \approx \sqrt{\frac{y}{2n}} - \frac{1}{6n} \qquad [3.15]$$

where $y = -\ln(\alpha/2)$. This formula is reasonably exact for fairly small values of $n$. For example, the approximation of the critical value $D_{0.05}(20) = 0.2953$ (see [BOL 83], Table 6.2) is $D_{0.05}(20) \approx 0.29535$ (first [3.15] approximation), $D_{0.05}(20) \approx 0.29403$ (second [3.15] approximation).

**One-sided alternatives.** If the alternatives are one-sided, i.e.

$$\bar{H}_1 : \sup_{x \in \mathbf{R}} (F(x) - F_0(x)) > 0, \ F(x) \geq F_0(x), x \in \mathbf{R}$$

or

$$\bar{H}_2 : \sup_{x \in \mathbf{R}} (F_0(x) - F(x)) > 0, \ F(x) \leq F_0(x), x \in \mathbf{R}$$

then the test is based on the statistics $D_n^+$ or $D_n^-$. Both have identical distributions because of symmetry. The hypothesis $H_0$ is rejected if

$$D_n^+ > D_\alpha^+(n) \quad \text{or} \quad D_n^- > D_\alpha^+(n)$$

respectively; here $D_\alpha^+(n)$ is the $\alpha$-critical value of the statistic $D_n^+$.

The cdf of the statistic $D_n^+$ (see [SMI 39b]) for all $x \in [0, 1)$ is

$$F_{D_n^+}(x) = \mathbf{P}\{D_n^+ \leq x\} = 1 - (1 - x)^n -$$

$$x \sum_{j=1}^{[n(1-x)]} C_n^j \left(1 - x - \frac{j}{n}\right)^{n-j} \left(x + \frac{j}{n}\right)^{j-1} \qquad [3.16]$$

The $P$-values under the alternatives $\bar{H}_1$ and $\bar{H}_2$ are $pv = 1 - F_{D_n^+}(D_n^+)$ and $pv = 1 - F_{D_n^-}(D_n^-)$, respectively. By the weak invariance property of the empirical process (see Appendix B, Theorem B.1.)

$$\sqrt{n}D_n^+ \xrightarrow{d} \sup_{0 \leq t \leq 1} B(t)$$

The proof of property 4 of the Brownian motion (Appendix B, section B.6) implies that for all $x > 0$

$$\mathbf{P}\{\sqrt{n}D_n^+ \le x\} \to K^+(x) = 1 - e^{-2x^2} \quad \text{as} \quad n \to \infty \qquad [3.17]$$

so the asymptotic $P$-value under the alternatives $\bar{H}_1$ and $\bar{H}_2$ are $pv_a = 1 - K^+(\sqrt{n}D_n^+)$ and $pv_a = 1 - K^+(-\sqrt{n}D_n^-)$, respectively.

**Example 3.1.** Powder is packed into bags. The weight of each packet should be 1 kg. The weights $X_i$ of 20 bags are given in the following table.

| $i$ | 1 | 2 | 3 | 4 | 5 | 6 | 7 |
|---|---|---|---|---|---|---|---|
| $X_i$ | 0.9473 | 0.9655 | 0.9703 | 0.9757 | 0.9775 | 0.9788 | 0.9861 |
| $i$ | 8 | 9 | 10 | 11 | 12 | 13 | 14 |
| $X_i$ | 0.9887 | 0.9964 | 0.9974 | 1.0002 | 1.0016 | 1.0077 | 1.0084 |
| $i$ | 15 | 16 | 17 | 18 | 19 | 20 | |
| $X_i$ | 1.0102 | 1.0132 | 1.0182 | 1.0225 | 1.0248 | 1.0306 | |

**Table 3.1.** *The data*

Test the hypothesis that the bag weight is normally distributed with the mean $\mu = 1$ kg and the standard deviation $\sigma = 25g$. We have $D_n = 0.1106$. The asymptotic $P$-value is $pv_a = 1 - K(\sqrt{20}\, 0.1106) = 0.9673$. The data do not contradict the hypothesis.

## 3.3. $\omega^2$, Cramér–von-Mises and Andersen–Darling tests

It was noted above that the $\omega^2$, Cramér–von-Mises and Andersen–Darling tests for the simple hypothesis $H_0 : F(x) \equiv F_0(x)$ are based on the statistics

$$\omega_n^2 = \int_{-\infty}^{\infty} (\hat{F}_n(x) - F_0(x))^2 \psi(F_0(x)) dF_0(x) = \int_0^1 (\hat{G}_n(y) - y)^2 \psi(y) dy$$

$$C_n = \int_{-\infty}^{\infty} (\hat{F}_n(x) - F_0(x))^2 dF_0(x) = \int_0^1 (\hat{G}_n(y) - y)^2 dy$$

$$A_n = \int_0^1 \frac{(\hat{F}_n(x) - F_0(x))^2}{F_0(x)(1 - F_0(x))} dF_0(x) = \int_0^1 \frac{(\hat{G}_n(y) - y)^2}{y(1 - y)} dy \quad [3.18]$$

where $\psi(t)$ is a non-negative function defined on the interval $[0, 1]$.

Deviation from the hypothesis is measured using quadratic metrics with the weight

$$\bar{H} : \int_{-\infty}^{\infty} (F(x) - F_0(x))^2 \psi(F_0(x)) dF_0(x) > 0 \qquad [3.19]$$

Cramér–von-Mises and Andersen–Darling tests are directed against alternatives with $\psi(y) = 1$ and $\psi(y) = 1/(y(1 - y))$, respectively.

**Computing the $\omega^2$, Cramér–von-Mises and Andersen–Darling statistics.** Let $\psi(y)$, $t\psi(y)$, $y^2\psi(y)$ be an integrable functions on $[0, 1]$. Set $Y_i = F_0(X_{(i)})$, $i = 1, \ldots, n$, $Y_{(0)} = 0$, $Y_{(n+1)} = 1$

$$g(y) = \int_0^y x\psi(x)dx, \quad h(y) = \int_0^y \psi(x)dx, \quad k = \int_0^1 (1 - y)^2 \psi(y)dy$$

**Theorem 3.4.** $\omega^2$, *Cramér–von-Mises and Andersen–Darling statistics can be written in the following way:*

$$\omega_n^2 = \frac{2}{n} \sum_{i=1}^n [g(Y_{(i)}) - \frac{2i - 1}{2n} h(Y_{(i)})] + k \qquad [3.20]$$

$$nC_n = \frac{1}{12n} + \sum_{i=1}^n \left( Y_{(i)} - \frac{2i - 1}{2n} \right)^2 \qquad [3.21]$$

$$nA_n = -n - \frac{1}{n} \left[ \sum_{i=1}^n (2i - 1)[\ln Y_{(i)} + \ln(1 - Y_{(n-i+1)})] \right] \qquad [3.22]$$

**Proof.** Using [3.18] we obtain

$$\omega_n^2 = \sum_{i=0}^{n} \int_{Y_{(i)}}^{Y_{(i+1)}} (\frac{i}{n} - y)^2 \psi(y) dy = \sum_{i=0}^{n} \frac{i^2}{n^2}[h(Y_{(i+1)}) - h(Y_{(i)})]-$$

$$2\sum_{i=0}^{n} \frac{i}{n}[g(Y_{(i+1)}) - g(Y_{(i)})] + \int_0^1 y^2 \psi(y) dy =$$

$$\sum_{i=1}^{n} \frac{(i-1)^2 - i^2}{n^2} h(Y_{(i)}) + h(1) + 2\sum_{i=1}^{n} \frac{i}{n} g(Y_{(i)}) - 2g(1)+$$

$$\int_0^1 y^2 \psi(y) dy =$$

$$= \frac{2}{n} \sum_{i=1}^{n} [g(Y_{(i)}) - \frac{2i-1}{2n} h(Y_{(i)})] + k$$

If $\psi(y) \equiv 1$ then [3.18] is the Cramér–von-Mises statistic. We obtain

$$g(y) = \frac{y^2}{2}, \quad h(y) = y, \quad k = \frac{1}{3}$$

$$C_n = \frac{1}{n} \sum_{i=1}^{n} \left( Y_{(i)}^2 - \frac{2i-1}{n} Y_{(i)} \right) + \frac{1}{3} =$$

$$\frac{1}{n} \sum_{i=1}^{n} \left( Y_{(i)}^2 - 2\frac{2i-1}{2n} Y_{(i)} + \left(\frac{2i-1}{2n}\right)^2 \right) -$$

$$\frac{1}{n} \sum_{i=1}^{n} \left(\frac{2i-1}{2n}\right)^2 + \frac{1}{3} = \frac{1}{12n^2} + \frac{1}{n} \sum_{i=1}^{n} \left( Y_{(i)} - \frac{2i-1}{2n}\right)^2$$

If $\psi(y) = 1/(y(1-y))$ then the functions $\psi(y)$, $t\psi(y)$, $y^2\psi(y)$ are not integrable on $[0, 1]$, so we can not define the Andersen–Darling statistic $A_n$ using [3.20]. Fix $0 < \varepsilon < Y_{(1)}$, $0 < \delta < 1 - Y_{(n)}$ and define the statistic $A_n$ as the limit

$$A_n = \lim_{\varepsilon,\delta \to 0} \int_\varepsilon^{1-\delta} (\hat{G}_n(y) - y)^2 \frac{dy}{y(1-y)}$$

Then, instead of $g(y), h(y)$ and $k$, write

$$g(y; \varepsilon, \delta) = \ln(1 - \varepsilon) - \ln(1 - y), \quad h(y; \varepsilon, \delta) =$$

$$\ln y - \ln \varepsilon + \ln(1 - \varepsilon) - \ln(1 - y)$$

and

$$k(\varepsilon, \delta) = \ln(1 - \delta) - \ln \varepsilon - 1 + \delta + \varepsilon, \quad \varepsilon \le t \le 1 - \delta$$

in [3.20]. We obtain

$$A_n = -\frac{2}{n} \sum_{i=1}^{n} \left[ \frac{2i-1}{2n} \ln Y_{(i)} + (1 - \frac{2i-1}{2n}) \ln(1 - Y_{(i)}) \right] +$$

$$+ \lim_{\varepsilon, \delta \to 0} \{\ln(1 - \delta) - \ln(1 - \varepsilon) - 1 + \delta + \varepsilon\}$$

$$nA_n = -n - 2 \sum_{i=1}^{n} \left[ \frac{2i-1}{2n} \ln Y_{(i)} + (1 - \frac{2i-1}{2n}) \ln(1 - Y_{(i)}) \right] =$$

$$-n - \frac{1}{n} \sum_{i=1}^{n} (2i - 1) \left[ \ln Y_{(i)} + \ln(1 - Y_{(n-i+1)}) \right]$$

$\triangle$

$\omega^2$, **Cramér–von-Mises and Andersen–Darling tests:** the hypothesis $H_0$ is rejected if $\omega^2$, $nC_n$ or $nA_n$ exceed their critical values.

For small $n$ the critical values are computed by statistical software (SAS, for example).

If $n$ is large then the critical values are found using asymptotic results by [3.7]

$$nC_n \xrightarrow{d} C = \int_0^1 B^2(t)dt, \quad nA_n \xrightarrow{d} A = \int_0^1 \frac{B^2(t)}{t(1-t)} dt$$

Using the fact that (see [MAR 77])

$$B(x) = \frac{\sqrt{2}}{\pi} \sum_{k=1}^{\infty} \frac{\sin \pi kx}{k} Z_k$$

where $Z_1, Z_2, \ldots$ is the sequence of iid random variables, $Z_i \sim N(0, 1)$, we have

$$C = \frac{1}{\pi^2} \sum_{k=1}^{\infty} \frac{Z_k^2}{k^2}, \quad A = \sum_{k=1}^{\infty} \frac{Z_k^2}{k(k+1)}$$

The cdfs of the random variables $C$ and $A$ are (see [BOL 83])

$$\mathbf{P}\{C \leq x\} = a_1(x) = 1 - \frac{1}{\pi} \sum_{j=1}^{\infty} (-1)^{j+1} \int_{(2j-1)^2\pi^2}^{4j^2\pi^2} \sqrt{\frac{-\sqrt{y}}{\sin(\sqrt{y})}} \frac{e^{-xy/2}}{y} dy$$

[3.23]

$$\mathbf{P}\{A \leq x\} = a_2(x) = \frac{\sqrt{2\pi}}{x} \sum_{j=0}^{\infty} (-1)^j \frac{j\Gamma(j+1/2)(4j+1)}{\Gamma(1/2)\Gamma(j+1)} \times$$

$$\int_0^{\infty} \exp\{\frac{x}{8(y^2+1)} - \frac{(4j+1)^2\pi^2(1+y^2)}{8x}\} dy$$

[3.24]

**Example 3.2.** (continuation of Example 3.1). Let us solve the problem in Example 3.1 using Cramér–von-Mises and Andersen–Darling tests. We obtain

$$nC_n = \frac{1}{240} + \sum_{i=1}^{20} (\Phi(z_i) - \frac{2i-1}{40})^2 = 0.0526, \quad z_i = \frac{X_i - 1}{0.025}$$

$$nA_n = -20 - 2\sum_{i=1}^{20} \left( \frac{2i-1}{40} \ln \Phi(z_i) + (1 - \frac{2i-1}{40}) \ln(1 - \Phi(z_i)) \right)$$

$$= 0.3971$$

The asymptotic $P$-values for the CM and AD tests are

$$pv_a = \mathbf{P}\{nC_n > 0.0536\} \approx 1 - a_1(0.0526) \approx 0.86$$

and

$$pv_a = \mathbf{P}\{nA_n > 0.3971\} \approx 1 - a_2(0.4026) \approx 0.85$$

respectively.

The $P$-values of the Andersen–Darling and Cramér–von-Mises tests practically coincide and the data do not contradict the hypotheses. The $P$-values of these tests are considerably different from the $P$-values of the Kolmogorov–Smirnov test.

**Comment 3.1.** In [LEM 04, LEM 09a, LEM 09b] the powers of various goodness-of-fit tests were investigated by simulation. In most cases the tests may be written in decreasing power order as follows: *Pearson chi-squared* ≻ *Andersen–Darling* ≻ *Cramér–von-Mises* ≻ *Kolmogorov–Smirnov*.

### 3.4. Modifications of Kolmogorov–Smirnov, Cramér–von-Mises and Andersen–Darling tests: composite hypotheses

In practice, goodness-of-fit tests for composite hypotheses are usually needed.

#### Composite hypothesis

$$H_0 : F(x) \in \mathcal{F}_0 = \{F_0(x; \boldsymbol{\theta}), \boldsymbol{\theta} \in \Theta \subset \mathbf{R}^m\}$$

The cdf $F_0(x; \boldsymbol{\theta})$ depends on the unknown finite-dimensional parameter $\boldsymbol{\theta}$.

#### Idea of test construction

The modified Kolmogorov–Smirnov, $\omega^2$, Cramér–von-Mises and Andersen–Darling tests are based on the statistics

$$D_n^{(mod)} = \sup_{y \in [0, 1]} |\hat{G}_n(y) - y|, \quad \omega_n^{2(mod)} = \int_0^1 (\hat{G}_n(y) - y)^2 \psi(y) dy$$

$$C_n^{(mod)} = \int_0^1 (\hat{G}_n(y) - y)^2 dy, \quad A_n^{(mod)} = \int_0^1 \frac{(\hat{G}_n(y) - y)^2}{y(1-y)} dy$$

$$[3.25]$$

where

$$\hat{G}_n(y) = \frac{1}{n} \sum_{i=1}^n \mathbf{1}_{(-\infty, y]}(Y_i) \qquad [3.26]$$

is the empirical distribution function of the random variables $Y_i = F_0(X_i, \hat{\theta})$; here $\hat{\theta}$ is the ML (or other) estimator of $\theta$.

Note that these statistics are natural modifications of the statistics $D_n, \omega_n^2, C_n$ and $A_n$: the last are written in the same form (see Theorem 3.1), the only difference being that in expressions of the modified statistics (composite hypothesis) $\hat{G}_n(y)$ means the empirical distribution function of the random variables $Y_i = F_0(X_i, \hat{\theta})$, whereas in the expressions of non-modified statistics (simple hypothesis) it means the empirical distribution function of the random variables $Y_i = F_0(X_i)$ with the completely known cdf $F_0$.

In general, the distribution of the modified statistics depends on the cumulative distribution function $F_0(x; \theta)$, but for some families of distributions it does not depend on unknown parameters.

**Theorem 3.5.** *Suppose that the ML estimators of unknown parameters are used. Then for location-scale families of distributions, i.e. for the families*

$$\{F_0(x; \theta) = G((x - \mu)/\sigma), \ \mu \in \mathbf{R}, \sigma > 0\}$$

*and shape-scale families of distributions, i.e. for the families*

$$\{F_0(x; \theta) = G((\frac{x}{\theta})^\nu), \ \theta > 0, \nu > 0\}$$

*the distribution of the modified statistics does not depend on unknown parameters but is different for different $G$.*

**Proof.** Let us consider a specified (i.e. $G$ is specified) location-scale family of distributions. The likelihood and loglikelihood functions are

$$L(\mu, \sigma) = \prod_{i=1}^{n} \frac{1}{\sigma} g(\frac{X_i - \mu}{\sigma}), \quad l(\mu, \sigma) = -n \ln \sigma + \sum_{i=1}^{n} h(\frac{X_i - \mu}{\sigma}),$$

$$g = G', \ h = \ln g$$

So the ML estimators $\hat{\mu}$ and $\hat{\sigma}$ of the parameters $\mu$ and $\sigma$ verify the system of equations

$$n + \sum_{i=1}^{n} \frac{X_i - \hat{\mu}}{\hat{\sigma}} h'(\frac{X_i - \hat{\mu}}{\hat{\sigma}}) = 0, \quad \sum_{i=1}^{n} h'(\frac{X_i - \hat{\mu}}{\hat{\sigma}}) = 0 \quad [3.27]$$

The cdf of each of the random variables $Z_i = (X_i - \mu)/\sigma$ is $G$ and does not depend on $\mu$ and $\sigma$. Using the relation $X_i = \sigma Z_i + \mu$, write the system of equations [3.27] in the form

$$n + \sum_{i=1}^{n} (\frac{\sigma}{\hat{\sigma}} Z_i + \frac{\mu - \hat{\mu}}{\hat{\sigma}}) h'(\frac{\sigma}{\hat{\sigma}} Z_i + \frac{\mu - \hat{\mu}}{\hat{\sigma}}) = 0$$

$$\sum_{i=1}^{n} h'(\frac{\sigma}{\hat{\sigma}} Z_i + \frac{\mu - \hat{\mu}}{\hat{\sigma}}) = 0$$

So the distribution of the random variables $\sigma/\hat{\sigma}$ and $(\mu - \hat{\mu})/\hat{\sigma}$ does not depend on unknown parameters. Using the equality

$$Y_i = \frac{X_i - \hat{\mu}}{\hat{\sigma}} = \frac{X_i - \mu}{\sigma} \frac{\sigma}{\hat{\sigma}} + \frac{\mu - \hat{\mu}}{\hat{\sigma}}$$

we obtain the distribution of $Y_i$, so the distributions of $\hat{G}_n(y)$ and the modified Kolmogorov–Smirnov and $\omega^2$ statistics do not depend on unknown parameters. The case of shape-scale families is considered analogously.

$\triangle$

**Comment 3.2.** Theorem 3.5 implies that the modified tests for exponential, normal, logistic, extreme values (scale or location-scale families), or lognormal, Weibull, loglogistic (shape-scale families) are parameter free. Tests for shape-scale families can be used in two ways: directly, or using tests for location-scale families, because the distribution of the logarithms of the random variables from such families belongs to location-scale families of distributions. It is sufficient to transform the data by taking the logarithms of $X_i$.

**Comment 3.3.** Even if $n$ is large, the critical values of the modified test may be very different from the critical values of the standard tests. For example, the 0.01-critical value of the modified Kolmogorov–Smirnov test (normal distribution) is near the 0.2-critical value of the standard test if $n$ is large. The more parameters are estimated, the greater the difference. So, if the composite hypothesis is tested then we cannot use the critical values of standard test statistics. The use of these values often leads to the conclusion that the data do not contradict the zero hypothesis, whereas, if we use the correct critical values of the modified statistics, the zero hypothesis might be rejected.

Some statistical packages such as SAS contain critical values for the modified statistic. Many tables of critical values for modified tests can be found in [DAG 86].

**Modified Kolmogorov–Smirnov, $\omega^2$, Cramér–von-Mises and Andersen–Darling tests for location-scale and shape-scale families:** the composite hypothesis $H_0$ is rejected if $D_n^{(mod)}$, $\omega_n^{2(mod)}$, $C_n^{(mod)}$ or $A_n^{(mod)}$ exceed their critical values.

If $n$ is large then an asymptotic distribution of the modified test statistics is used to approximate the $P$-values of the tests.

## Asymptotic distribution of the modified test statistics

Since all modified statistics are functionals of the empirical process

$$\xi_n(y) = \sqrt{n}(\hat{G}_n(y) - y), \ y \in (0,1) \qquad [3.28]$$

their limit distribution can be found if the limit distribution of the empirical process $\xi_n(y)$ is found.

The asymptotic distributions of the process $\xi_n(y)$ for location-scale and shape-scale families are as follows (see [MAR 77]): $\xi_n(y) \xrightarrow{d} \xi(x)$ on $(0,1)$, where $\xi(y)$ is a Gaussian process with zero mean and the correlation function

$$K(x,y) = x \wedge y - xy - \frac{1}{a}K_1(x,y)$$

$$K_1(x,y) = c_2 w_1(x)w_2(y) + c_1 w_2(x)w_2(y) - c_3[w_1(x)w_2(y) + w_2(x)w_1(y)]$$

where

a) *location-scale family*

$$w_1(x) = g(G^{-1}(x)), \quad w_2(x) = G^{-1}(x)g(G^{-1}(x))$$

$$c_1 = \int_{-\infty}^{\infty} \frac{(g'(x))^2}{g(x)}dx, \quad c_2 = \int_{-\infty}^{\infty} x^2\frac{(g'(x))^2}{g(x)}dx - 1$$

$$c_3 = \int_{-\infty}^{\infty} x\frac{(g'(x))^2}{g(x)}dx, \quad a = c_1c_2 - c_3^2$$

b) *shape-scale family*

$$w_1(x) = G^{-1}(x)g(G^{-1}(x)), \quad w_2(x) = G^{-1}(x)g(G^{-1}(x))\ln G^{-1}(x)$$

$$c_1 = \int_{-\infty}^{\infty} \left(1 + x\frac{g'(x)}{g(x)}\right)^2 g(x)dx,$$

$$c_2 = \int_{-\infty}^{\infty} \left(1 + x\ln x\frac{g'(x)}{g(x)} + \ln x\right)^2 g(x)dx$$

$$c_3 = \int_{-\infty}^{\infty} \left(1 + x\frac{g'(x)}{g(x)}\right)\left(1 + x\ln x\,\frac{g'(x)}{g(x)} + \ln x\right)g(x)dx$$

$$a = c_1 c_2 - c_3^2$$

By the weak invariance property (see Appendix B, Theorem B.1) the modified Kolmogorov–Smirnov, Cramér–von-Mises and Andersen–Darling statistics have the following limits

$$\sqrt{n}D_n^{(mod)} \xrightarrow{d} \sup_{0\le t\le 1} |\xi(t)|, \quad nC_n^{(mod)} \xrightarrow{d} \int_0^1 \xi^2(t)dt$$

$$nA_n^{(mod)} \xrightarrow{d} \int_0^1 \frac{\xi^2(t)}{t(1-t)}dt \qquad\qquad [3.29]$$

Some statistical packages such as SAS contain asymptotic critical values of the modified statistic.

**Example 3.3** (continuation of Example 2.5). Using the data in Example 2.5, verify the same hypotheses: the random variables $V_i$ are a) normally, b) lognormally distributed.

a) Compute the modified Kolmogorov–Smirnov, Cramér–von-Mises and Andersen–Darling statistics:

$$\bar{X} = 12.0184, \quad s = 10.0325, \quad D_{49}^{(mod)+} = 0.1806$$

$$D_{49}^{(mod)-} = 0.1296, \quad D_{49}^{(mod)} = 0.1806$$

$$nC_n = \frac{1}{588} + \sum_{i=1}^{49}(\Phi(Y_{(i)}) - \frac{2i-1}{98})^2 = 0.3241$$

$$Y_{(i)} = \frac{X_{(i)} - 12.0184}{10.03248}$$

$$nA_n = -49 - 2\sum_{i=1}^{49}\left(\frac{2i-1}{49}\ln \Phi(Y_{(i)}) + (1 - \frac{2i-1}{49})\ln(1 - \Phi(Y_{(i)}))\right)$$

$$= 1.8994$$

The $P$-values computed using the SAS package are

$$pv^{KSmod} < 0.01, \quad pv^{CMmod} < 0.005, \quad pv^{ADmod} < 0.005$$

The normality hypothesis is rejected.

If the $P$-values are computed using the same values of the modified test statistics but the probability distribution of the non-modified standard KS, CM and AD statistics (which is very common in many books and statistical packages) then the computed $P$-values are much larger: 0.0724, 0.1193 and 0.1049, respectively. For example, using the SPSS package, we obtained a $P$-value of 0.0724 in the Kolmogorov test for a composite hypothesis. We can see that incorrectly used tests do not give contradictions to the zero hypothesis (if the significance levels are smaller than 0.0724, 0.1193 and 0.1049, for the KS, CM and AD tests, respectively).

b) Using the logarithms of the initial observations we compute the values of the modified Kolmogorov–Smirnov, Cramér–von-Mises and Andersen–Darling statistics

$$\bar{X} = 2.1029, \quad s = 0.9675, \quad D_{49}^{(mod)} = 0.1033, \quad nC_n = 0.0793,$$

$$nA_n = 0.5505$$

The $P$-values computed using the SAS package are

$$pv^{KSmod} = 0.6723, \quad pv^{CMmod} = 0.2141, \quad pv^{ADmod} = 0.1517$$

The data do not contradict the hypothesis that $\ln V$ is normally distributed. Incorrectly used standard tests do not reject the hypothesis either.

**Comment 3.5.** In [LEM 04, LEM 09a, LEM 09b] the power of various goodness-of-fit tests in testing various composite hypotheses was investigated by simulation. For the majority of hypotheses and alternatives the tests may be written in decreasing power order as follows: *modified Andersen–Darling* ≻ *modified chi-squared* ≻ *modified Cramér–von-Mises* ≻ *Pearson chi-squared* ≻ *modified Kolmogorov–Smirnov*.

## 3.5. Two-sample tests

### 3.5.1. *Two-sample Kolmogorov–Smirnov tests*

Suppose that two independent simple samples

$$\mathbf{X} = (X_1, ..., X_m)^T \quad \text{and} \quad \mathbf{Y} = (Y_1, ..., Y_n)^T$$

are obtained, observing absolutely continuous independent random variables $X$ and $Y$ with cumulative distribution functions $F_1$ and $F_2$, respectively. Denote by $\hat{F}_{1m}(x)$ and $\hat{F}_{2n}(x)$ the empirical distribution functions obtained from the samples $\mathbf{X}$ and $\mathbf{Y}$, respectively.

Let us consider the homogeneity hypothesis

$$H_0 : F_1(x) = F_2(x), \;\; \forall x \in \mathbf{R} \qquad\qquad [3.30]$$

against the two-sided alternative

$$\bar{H} : \sup_{x \in \mathbf{R}} |F_1(x) - F_2(x)| > 0 \qquad\qquad [3.31]$$

or the one-sided alternatives

$$\bar{H}^+ : \sup_{x \in \mathbf{R}} (F_1(x) - F_2(x)) > 0, \; F_1(x) \geq F_2(x), x \in \mathbf{R}$$

$$\bar{H}^- : \inf_{x \in \mathbf{R}} (F_1(x) - F_2(x)) < 0, \; F_1(x) \leq F_2(x), x \in \mathbf{R} \qquad [3.32]$$

*Two-sample Kolmogorov–Smirnov tests* for the hypothesis $H_0$ against the two-sided alternative $\bar{H}$ are based on the statistic

$$D_{m,n} = \sup_{|x| < \infty} |\hat{F}_{1m}(x) - \hat{F}_{2n}(x)| \qquad\qquad [3.33]$$

As in Theorem 3.1, it can be shown that the distribution of the statistics $D_{m,n}$ under $H_0$ does not depend on the cumulative distribution function of observed random variables but depends on the sample sizes $m$ and $n$:

$$D_{m,n} = \sup_{x \in \mathbf{R}} |\hat{F}_{1m}(x) - \hat{F}_{2n}(x)| = \sup_{0 \leq y \leq 1} |\hat{G}_{1m}(y) - \hat{G}_{2n}(y)|$$

where $\hat{G}_{1m}$ and $\hat{G}_{2n}$ are the empirical distribution functions from the samples

$$(U_{11}, ..., U_{1m})^T \quad \text{and} \quad (U_{21}, ..., U_{2n})^T$$

obtained by observing random variables $U_1$ and $U_2$ independent uniformly distributed in $[0, 1]$; where

$$U_{1i} = F_1(X_i), \quad U_{2j} = F_2(Y_j)$$

In the case of one-sided alternatives tests are based on the following statistics

$$D_{m,n}^+ = \sup_{x \in \mathbf{R}} (\hat{F}_{1m}(x) - \hat{F}_{2n}(x)), \ D_{m,n}^- = - \inf_{x \in \mathbf{R}} (\hat{F}_{1m}(x) - \hat{F}_{2n}(x))$$

[3.34]

Since the functions $\hat{F}_{1m}(x)$ and $\hat{F}_{2n}(x)$ are step functions, the supremum is equal to the maximum deviation at the jump points of one of these functions. Analogously to the case of the one-sample Kolmogorov–Smirnov statistic, the statistic $D_{m,n}$ can be written as follows:

$$D_{m,n} = \max(D_{m,n}^+, \ D_{m,n}^-)$$

$$D_{m,n}^+ = \max_{1 \leq r \leq m} \left( \frac{r}{m} - \hat{F}_{2n}(X_{(r)}) \right) = \max_{1 \leq s \leq n} \left( \hat{F}_{1m}(Y_{(s)}) - \frac{s-1}{n} \right)$$

$$D_{m,n}^- = \max_{1 \leq r \leq m} \left( \hat{F}_{2n}(X_{(r)}) - \frac{r-1}{m} \right) = \max_{1 \leq s \leq n} \left( \frac{s}{n} - \hat{F}_{1m}(Y_{(s)}) \right)$$

**Two-sample Kolmogorov–Smirnov test:** the hypothesis $H_0$ is rejected with a significance level $\alpha$ if

$$D_{m,n} \geq D_\alpha(m, \ n)$$

where $D_\alpha(m, n)$ is the $\alpha$ critical value of $D_{m,n}$

$$\mathbf{P}\{D_{m,n} \geq D_\alpha(m, n)\} \leq \alpha, \quad \mathbf{P}\{D_{m,n} < D_\alpha(m, n)\} > 1 - \alpha$$

**Comment 3.6.** The statistic $D_{m,n}$ takes values of the form $l/k$; here $k = k(m, n)$ is the least common multiple of the numbers $m$ and $n$. To obtain the exact test significance level $\alpha$, the test is randomized. As was noted above, in practice a non-randomized test is used, taking a slightly smaller significance level $\alpha$.

To be precise, the critical value $D_\alpha(m, n)$ is understood as the least number of the form $l/k$ such that

$$\mathbf{P}\{D_{m, n} > D_\alpha(m, n)|H\} = \alpha' \leq \alpha$$

The critical value $D_\alpha(m,n)$ can be found using the approximation (see [BOL 83])

$$D_\alpha(m,n) \approx \frac{1}{k(m,n)} + D_\alpha(\nu) + \frac{1}{\nu} -$$

$$\frac{1}{\nu}\left[\frac{n-m}{6(m+n)} + \frac{1}{2}\frac{m-d(m,n)}{m+n+d(m,n)}\right]$$

where $k(m, n)$ and $d(m, n)$ are the least common multiple and the greatest common factor of $m$ and $n$, $\nu = mn/(m+n)$, $m \leq n$.

In the case of one-sided alternatives, the hypothesis $H_0$ is rejected if

$$D_{m,n}^+ > D_{2\alpha}(m, n) \text{ or } D_{m,n}^- > D_{2\alpha}(m, n) \qquad [3.35]$$

respectively. These tests are approximate but the approximation is good enough if $\alpha < 0.1$ (see [BOL 83]).

For small samples the critical values $D_\alpha(m, n)$ are given by most statistical software.

If $m$ and $n$ are large then the approximate critical values of the statistics $D_{m,n}^+$ and $D_{m,n}$ are found using an asymptotic distribution given in the following theorem.

**Theorem 3.6.** *Suppose that* $m/(m+n) \to p \in (0, 1)$, $m, n \to \infty$. *If the hypothesis* $H_0$ *holds then*

$$\mathbf{P}\{\sqrt{\frac{mn}{m+n}} D_{m,n} \leq x\} \to 1 - 2\sum_{n=1}^{\infty}(-1)^{n-1}e^{-2n^2x^2}, \quad as \quad n \to \infty$$
[3.36]

$$\mathbf{P}\left\{\sqrt{\frac{mn}{m+n}} D_{m,n}^+ \leq x\right\} \to 1 - e^{-2x^2}$$
[3.37]

**Proof.** Under the hypothesis $H_0$, Theorem 3.2 implies

$$\sqrt{\frac{mn}{m+n}}(\hat{F}_{1m} - \hat{F}_{2n}) =$$

$$\sqrt{\frac{n}{m+n}} \sqrt{m}\,[\hat{F}_{1m} - F_1] - \sqrt{\frac{m}{m+n}} \sqrt{n}\,[\hat{F}_{2n} - F_2] \overset{d}{\to} B =$$

$$\sqrt{1-p}\,B_1 - \sqrt{p}\,B_2;$$

where $B_1$ and $B_2$ are independent Brownian bridges. The stochastic process $B$ is Gaussian because it is a linear function of a Gaussian process. Moreover,

$$\mathbf{E}B(t) = 0, \quad \mathbf{cov}(B(s), B(t)) = (1 - \gamma)\mathbf{cov}(B_1(s), B_1(t)) +$$

$$\gamma\mathbf{cov}(B_2(s), B_2(t)) = s \wedge t - st$$

so $B$ is also a Brownian bridge. Hence

$$\sqrt{\frac{mn}{m+n}} D_{m,n} \overset{d}{\to} \sup_{0 \leq t \leq 1} |B(t)|, \quad \sqrt{\frac{mn}{m+n}} D_{m,n}^+ \overset{d}{\to} \sup_{0 \leq t \leq 1} B(t)$$

$\triangle$

The asymptotic $P$-value based on the last theorem is

$$pv_a = 1 - \mathbf{P}\{ \sup_{0 \leq t \leq 1} |B(t)| < \sqrt{\frac{mn}{m+n}} D_{m,n}\} = 1 - K(\sqrt{\frac{mn}{m+n}} D_{m,n})$$

where $K(x)$ is the cdf of the Kolmogorov distribution.

**Example 3.4.** The influence of fungicides on the number of coffee bean infections is investigated. The numbers of infected beans in two coffee-tree groups (as percentages) are given in the following table:

| Fung. used | 6.01 | 2.48 | 1.76 | 5.10 | 0.75 | 7.13 | 4.88 |
|---|---|---|---|---|---|---|---|
| Fung. not used | 5.68 | 5.68 | 16.30 | 21.46 | 11.63 | 44.20 | 33.30 |

Test the hypothesis that fungicides have no influence on the percentage of infections.

The computations are given in the following table:

| $r$ | $X_{(r)}$ | $Y_{(r)}$ | $\frac{r}{n}$ | $\hat{F}_{1n}(Y_{(r)})$ | $\hat{F}_{2n}(X_{(r)})$ | $\frac{r}{n} - \hat{F}_{1n}(Y_{(r)})$ | $\frac{r}{m} - \hat{F}_{2n}(X_{(r)})$ |
|---|---|---|---|---|---|---|---|
| 1 | 0.75 | 5.68 | 1/7 | 5/7 | 0 | $-4/7$ | 1/7 |
| 2 | 1.76 | 5.68 | 2/7 | 5/7 | 0 | $-3/7$ | 2/7 |
| 3 | 2.48 | 11,63 | 3/7 | 1 | 0 | $-4/7$ | 3/7 |
| 4 | 4.88 | 16.30 | 4/7 | 1 | 0 | $-3/7$ | 4/7 |
| 5 | 5,10 | 21,46 | 5/7 | 1 | 0 | $-2/7$ | 5/7 |
| 6 | 6.01 | 33,30 | 6/7 | 1 | 2/7 | $-1/7$ | 4/7 |
| 7 | 7.13 | 44,20 | 1 | 1 | 2/7 | 0 | 5/7 |

We have

$$D^+_{m,n} = \max_{1 \le r \le m} \left( \frac{r}{m} - \hat{F}_{2n}(X_{(r)}) \right) = \frac{5}{7}$$

$$D^-_{m,n} = \max_{1 \le r \le n} \left( \frac{r}{n} - \hat{F}_{1n}(Y_{(r)}) \right) = 0, \quad D_{m,n} = \frac{5}{7} \approx 0.714286$$

The SPSS package gives the $P$-value

$$pv = \mathbf{P}\{D_{m,n} \ge 0.714286\} = 0.05303$$

The asymptotic $P$-value is $pv_a = 0.05623$.

The data do not contradict the hypothesis if the significance level is 0.05 but reject the hypothesis if the significance level is 0.1. So larger samples or more powerful tests are needed to reach the final conclusion. We shall see below that the two-sample Cramér–von-Mises and Wilcoxon tests better detect the differences between the distributions of these two samples.

### 3.5.2. *Two-sample Cramér–von-Mises test*

The two-sample Cramér–von-Mises test is based on the statistic (see [AND 62])

$$T_{m,n} = \frac{mn}{m+n} \int_{-\infty}^{+\infty} (\hat{F}_{1m}(x) - \hat{F}_{2n}(x))^2 d\hat{G}_{m+n}(x) \qquad [3.38]$$

where

$$\hat{G}_{m+n}(x) = \frac{m}{m+n}\hat{F}_{1m}(x) + \frac{n}{m+n}\hat{F}_{2n}(x)$$

is the empirical distribution function obtained from a unified sample $(X_1, ..., X_m, Y_1, ..., Y_n)^T$ of size $m+n$.

Using definition [3.38], it can be proved (see exercise 3.3) that the statistic $T_{m,n}$ can be written in the form

$$T_{m,n} = \frac{1}{mn(m+n)}[m\sum_{j=1}^{m}(R_{1j}-j)^2 + n\sum_{i=1}^{n}(R_{2i}-i)^2] - \frac{4mn-1}{6(m+n)}$$
$$[3.39]$$

where $R_{1j}$ and $R_{2i}$ are the positions of the elements of the first and second sample, respectively, in the unified ordered sample.

Using the Rosenblatt's results [ROS 52] it can be shown that there is a limit distribution of the statistic $T_{m,n}$ as $m, n \to \infty$, $m/n \to \lambda$, $0 < \lambda < \infty$. This limit distribution coincides with the limit distribution of the statistic $nC_n$ given in [3.23]:

$$\mathbf{P}\{T_{m,n} \le x\} \to \mathbf{P}\{C \le x\} = a_1(x)$$

The mean and the variance of the statistic $C$ are $1/6$ and $1/45$ (see exercise 3.1), and the moments of the statistic $T_{m,n}$ are (see exercise 3.4)

$$\mathbf{E}T_{m,n} = \frac{1}{6}(1 + \frac{1}{m+n}),$$

$$\mathbf{Var}(T_{m,n}) = \frac{1}{45}(1 + \frac{1}{m+n})(\frac{m+n+1}{m+n} - \frac{3(m+n)}{4mn})$$

so instead of $T_{m,n}$, the modified statistic

$$T_{m,n}^* = \frac{T_{m,n} - \mathbf{E}T_{m,n}}{\sqrt{45\mathbf{Var}(T_{m,n})}} + \frac{1}{6} \qquad [3.40]$$

is used. The first two moments of this statistic coincide with the corresponding moments of the random variable $C$.

**Asymptotic Cramér–von-Mises two-sample test:** the homogeneity hypothesis is rejected if

$$T_{m,n}^* > t_\alpha^*(m, n) \qquad [3.41]$$

where $t_\alpha^*(m, n)$ is the critical value verifying the equality

$$1 - a_1(t_\alpha^*(m, n)) = \alpha$$

The approximation by the cdf $a_1(x)$ is sufficiently exact for relatively small $m, n$. An analysis of the accuracy of the approximation can be found in [BOL 83].

**Example 3.5.** (continuation of Example 3.4). Let us apply the Cramér–von-Mises test to the data considered in Example 3.4. We have $T_{7,7} = 0.7704$ and the value of the modified statistic is $T_{7,7}^* = 0.7842$. The asymptotic $P$-value $pv_a = 1 - a_1(0.7842) \approx 0.008$. The Cramér–von-Mises test much better detects the differences between the distributions of these two samples.

## 3.6. Bibliographic notes

Goodness-of-fit statistics for a simple hypothesis based on the supremum of the difference between empirical and theoretical distribution functions have been considered by

Von-Mises [MIS 31], Kolmogorov [KOL 33], Smirnov [SMI 44], Kuiper [KUI 60], Watson [WAT 76] and Darling [DAR 83a, DAR 83b].

Goodness-of-fit statistics for a simple hypothesis based on integral differences (Andersen–Darling, Cramér–von-Mises and other $\omega^2$-type statistics) have been considered by Smirnov [SMI 36, SMI 37, SMI 39b], Andersen and Darling [AND 52, AND 54], Darling [DAR 55, DAR 57] and Martynov [MAR 77].

The problem of modification of the Kolmogorov–Smirnov and $\omega^2$-type statistics for testing of composite hypotheses starts from the classical work by Kac *et al.* [KAC 55] which generated plenty of attention from researchers both in the theoretical and applied statistical literature: Lilliefors [LIL 67, LIL 69], Durbin [DUR 73], Tyurin [TYU 70, TYU 84], Stephens [STE 74, STE 76, STE 77, STE 79], Martynov [MAR 75, MAR 77], Khmaladze [KHM 77], D'Agostino and Stephens [DAG 86], Stute *et al.* [STU 93], Lemeshko *et al.* [LEM 98a, LEM 04, LEM 09a, LEM 09b, LEM 10] and Szuks [SZU 08].

Statistics based on the difference between two independent empirical distribution functions have been considered by Smirnov [SMI 39a], Watson [WAT 40], Lehmann [LEH 51], Andersen [AND 62], Maag and Stephens [MAA 68] and Pettitt [PET 76, PET 79].

Bakshaev [BAK 09, BAK 10] gives goodness-of-fit tests based on $N$-distances between non-parametric and parametric estimators of the cdf and homogeneity tests based on $N$-distances between empirical distribution functions. They can be used as goodness-of-fit and homogeneity tests not only in univariate but also in multivariate cases.

Nikitin [NIK 95] considered the asymptotic efficiency of various non-parametric tests.

## 3.7. Exercises

**3.1.** Find the first two moments of the random variables $C = \int_0^1 B^2(t)dt$ and $A = \int_0^1 B^2(t)/(t(1-t))dt$ given in formula [3.7].

**3.2.** Find the first two moments of the statistics $nC_n$ and $nA_n$ and compare them with the moments given in exercise 3.1.

**3.3.** Show that the Cramér–von-Mises two-sample statistic [3.37] can be written in the form [3.38].

**3.4.** Find the first two moments of the statistic $T_{m,n}$, defined by [3.37] and [3.38].

**3.5.** Construct a confidence zone for the cdf $F(x)$ of an absolutely continuous random variable. The confidence level is $Q = 1 - 2\alpha$.

**3.6.** Suppose that $X$ is a discrete random variable with possible values 0, 1, 2..., and the probabilities $p_k = \mathbf{P}\{X = k\}, k = 0, 1, \dots$. Prove that the random variable

$$Z = \sum_{k=0}^{X-1} p_k + p_X Y$$

is uniformly distributed in the interval $(0,\ 1)$ if the random variable $Y \sim U(0,\ 1)$ d does not depend on $X$.

**3.7.** (continuation of exercise 3.6). Suppose that $X_1, \dots, X_n$ is a simple sample of a discrete random variable $X$, and $\hat{F}_n(x)$ is the empirical distribution function. Verify the hypothesis $H\ :\ \mathbf{E}(\hat{F}_n(x))\ =\ F_0(x), |x|\ <\ \infty$; here $F_0(x)$ is the completely specified cumulative distribution function of a discrete random variable. Replace the hypothesis $H$ by the hypothesis $H'\ :\ \mathbf{E}(\hat{G}_n(z))\ =\ G(z), 0\ <\ z\ <\ 1$. Here $\hat{G}_n(z)$

is the empirical distribution function of the sample $Z_i = F_0(X_i) + p_{X_i} Y_k$, $i = 1, ..., n$, $k = 1, ..., n$; $Y_1, ..., Y_n$ – simple and independent on $X_1, ..., X_n$ sample of a random variable $Y \sim U(0, 1)$, $G(z)$ is the cumulative distribution function of a random variable $U(0, 1)$. We obtain a randomized analog of, for example, the Kolmogorov–Smirnov test for discrete distributions.

**3.8.** (continuation of exercise 3.7). Using the test given in exercise 3.7, solve exercise 2.5.

**3.9.** (continuation of exercise 3.7). Using the test given in exercise 3.7, solve exercise 2.7.

**3.10.** Apply the Kolmogorov–Smirnov, Andersen–Darling and Cramér–von-Mises tests to verify the hypothesis that the data in exercise 2.16 are obtained by observing a) a normal random variable with parameters which are values of $\hat{\mu}$ and $\hat{\sigma}^2$ obtained from these data, b) a normal random variable.

**3.11.** Apply the Kolmogorov–Smirnov, Andersen–Darling and Cramér–von-Mises tests to verify the hypothesis that the data in exercise 2.17 is obtained by observing a) a lognormal random variable with parameters which are the values of the estimators $\hat{\mu}$ and $\hat{\sigma}^2$, b) a lognormal random variable.

**3.12.** In the control of the stability of machine-tools, 20 products are inspected every hour. Using the measured values of a given parameter, the value of the unbiased estimator of the variance $s^2$ is found. The values are given in the table.

| | | | | | | | |
|---|---|---|---|---|---|---|---|
| 0.1225 | 0.1764 | 0.1024 | 0.1681 | 0.0841 | 0.0729 | 0.1444 | 0.0900 |
| 0.0961 | 0.1369 | 0.1521 | 0.1089 | 0.1296 | 0.1225 | 0.1156 | 0.1681 |
| 0.0676 | 0.0784 | 0.1024 | 0.1156 | 0.1024 | 0.0676 | 0.1225 | 0.1521 |
| 0.1369 | 0.1444 | 0.1521 | 0.1024 | 0.1089 | 0.1600 | 0.0961 | 0.1600 |
| 0.1024 | 0.1369 | 0.1089 | 0.1681 | 0.1296 | 0.1521 | 0.1600 | 0.0576 |
| 0.0784 | 0.1089 | 0.1156 | 0.1444 | 0.1296 | 0.1024 | 0.1369 | |

Apply the Kolmogorov–Smirnov, Andersen–Darling and Cramér–von-Mises tests to verify the hypothesis that the process was stable (in the sense of deviation of measured parameter values). Suppose that the measured parameter is normally distributed with the variance $\sigma^2 = 0.109$.

**3.13.** Generate samples of size 50 from a) a normal distribution $N(3,4)$; b) a lognormal distribution $LN(1,2)$; c) an Erlang distribution $G(3,4)$; and d) a Cauchy distribution $K(0,2)$. Apply the Kolmogorov–Smirnov, Andersen–Darling and Cramér–von-Mises tests to verify the respective hypotheses.

**3.14.** The results of two experiments with flies are given in the table. In the first experiment, the flies are in contact with a poison for 30 s, and in the second experiment for 60 s. The paralyzing effect of the poison is defined by the reaction time ($X_{1i}$ in the first and $X_{2i}$ in the second experiment), i.e. the time from contact with the poison to the moment when a fly cannot stand on its legs.

| $i$ | $X_{1i}$ | $i$ | $X_{1i}$ | $i$ | $X_{2i}$ | $i$ | $X_{2i}$ |
|---|---|---|---|---|---|---|---|
| 1 | 3.1 | 9 | 53.1 | 1 | 3,3 | 9 | 56.7 |
| 2 | 9.4 | 10 | 59.4 | 2 | 10.0 | 10 | 63,3 |
| 3 | 15.6 | 11 | 65.6 | 3 | 10.7 | 11 | 70.0 |
| 4 | 21.9 | 12 | 71.9 | 4 | 23.3 | 12 | 76.7 |
| 5 | 28,1 | 13 | 78,1 | 5 | 30.0 | 13 | 83,3 |
| 6 | 34,4 | 14 | 84,4 | 6 | 36.7 | 14 | 90.0 |
| 7 | 40.6 | 15 | 90.6 | 7 | 43,3 | 15 | 96.7 |
| 8 | 46,9 | 16 | 96,9 | 8 | 50.0 | | |

Apply the Kolmogorov–Smirnov and Cramér–von-Mises two- sample tests to verify the hypothesis that the elements of both samples have the same distribution.

**3.15.** Divide the data in exercise 2.16 into two samples (the first five and last five columns). Apply the Kolmogorov–Smirnov two-sample test to verify the hypothesis that the elements of both samples have the same distribution.

**3.16.** Divide the data in exercise 2.17 into two samples (the first five and last 10 columns). Apply the Kolmogorov–Smirnov and Cramér–von-Mises two-sample tests to verify the hypothesis that the elements of both samples have the same distribution.

## 3.8. Answers

**3.1.** $\mathbf{E}C = 1/6, \text{Var}C = 1/45; \mathbf{E}A = 1, \text{Var}A = 2\pi^2/3 - 6.$
**Note.** $\mathbf{E}C = \int_0^1 \mathbf{E}(B^2(t))dt, \mathbf{E}(C^2) = 2\int_0^1 \int_0^1 \mathbf{E}(B^2(t)B^2(s))dsdt.$
Use the fact that $B(t) \sim N(0, t(1-t)), 0 \le t \le 1; (B(s), B(t))^T \sim N_2(\mathbf{0}, \Sigma), \sigma_{11} = s(1-s), \sigma_{22} = t(1-t), \sigma_{12} = s(1-t), 0 \le s \le t \le 1.$

**3.2.** $\mathbf{E}(nC_n) = 1/6, \text{Var}(nC_n) = 1/45 - 1/(60n); \mathbf{E}(nA_n) = 1, \text{Var}(nA_n) = 2\pi^2/3 - 6 + (10 - \pi^2)/n.$ **Note.** $\mathbf{E}(nC_n) = \int_0^1 n\mathbf{E}(\hat{G}_n(y) - y)^2 dy, \mathbf{E}(nC_n)^2) = 2\int_0^1 \int_0^y n^2\mathbf{E}[(\hat{G}_n(x) - x)^2(\hat{G}_n(y) - y)^2]dxdy.$ The random variable $n\hat{G}_n(y) \sim B(n, y), 0 < y < 1;$ the random vector $(n\hat{G}_n(x), n(\hat{G}_n(y) - \hat{G}_n(x)), n(1 - \hat{G}_n(y)))^T \sim P_3(n, (x, y - x, 1 - y)), 0 \le x \le y \le 1.$ Find the expressions of the moments under the integral sign and integrate.

**3.5.** $\underline{F(x)} = \max(0, \hat{F}(x) - D_\alpha(n)), \overline{F(x)} = \min(\hat{F}(x) + D_\alpha(n), 1).$

**3.8.** $D_n = 0.0196, C_n = 0.0603, A_n = 0.3901.$ Respective $P$-values are $pv > 0.25.$ The data do not contradict the hypothesis.

**3.9.** $D_n = 0.0305, C_n = 0.0814, A_n = 0.5955$. Respective $P$-values are $pv > 0.25$. The data do not contradict the hypothesis.

**3.10.** $D_n = 0.0828, nC_n = 0.1139, nA_n = 0.7948$; a) asymptotic $P$=values are 0.5018, 0.5244, 0.4834. The data do not contradict the hypothesis. b) $P$-values are 0.0902, 0.0762, 0.0399. The hypothesis is rejected by all tests if the significance level is $\alpha = 0.1$.

**3.11.** $D_n = 0.0573, nC_n = 0.0464, nA_n = 0.3411$; asymptotic $P$-values are 0.6927, 0.8911, 0.8978. The data do not contradict the hypothesis. b) $P$-values are $pv > 0.15, pv > 0.25$, $pv > 0.25$. The data do not contradict the hypothesis.

**3.12.** $D_n = 0.2547, nC_n = 0.8637, nA_n = 4.3106$; asymptotic $P$=values are 0.0041, 0.0049, 0.0065. The hypothesis is rejected.

**3.14.** $D_{m,n} = 0.075$. The exact $P$-value is $pv = 0,99999995$, $pv_a = 0.999997$. The statistic $T_{m,n}$ takes the value 0.0114 and the modified statistic $T^*_{m,n}$ takes the value 0.0032. The asymptotic $P$-value is $pv_a = 1 - a_1(0.0032) = 0.999997$. The data do not contradict the hypothesis.

**3.15.** $D_{m,n} = 0.16, pv = 0.471, pv_a = 0.5441$. The statistic $T^*_{m,n}$ takes the value 0.2511. The asymptotic $P$-value is $pv_a = 1 - a_1(0.2511) = 0.187$. The data do not contradict the hypothesis.

**3.16.** $D_{m,n} = 0.13$. the exact $P$-value is $pv = 0,5227$, $pv_a = 0.6262$. The statistic $T^*_{m,n}$ takes the value 0.1668. The asymptotic $P$-value is $pv_a = 1 - a_1(0.0032) = 0.3425$. The data do not contradict the hypothesis.

# Chapter 4

# Rank Tests

## 4.1. Introduction

In previous chapters, we discussed two methods of non-parametric test construction: chi-squared type test statistics, which are functions of the numbers of sample elements falling into some sets, were discussed in Chapter 2; in Chapter 3 tests based on functionals of the difference between empirical and hypothetical theoretical cumulative distribution functions were discussed. The distribution of each test is obtained using the integral transformation (see Theorem 3.1), which transforms absolutely continuous random variables into random variables uniformly distributed on $[0, 1]$.

In this section we consider non-parametric tests based on statistics which depend only on the location of observations in the ordered sample and do not depend directly on their values.

## 4.2. Ranks and their properties

Suppose that $\mathbf{X} = (X_1, ..., X_n)^T$ is a simple sample of absolutely continuous random variable $X$, and $X_{(1)} < \cdots < X_{(n)}$ are the order statistics.

**Definition 4.1.** The rank $R_i$ of sample element $X_i$ is the order number of this element in the ordered sample $(X_{(1)}, ..., X_{(n)})$, i.e.

$$rank(X_i) = R_i = j \quad \text{if} \quad X_i = X_{(j)}$$

For example, if $(X_1, ..., X_5)^T = (63, 32, 41, 25, 38)^T$ then

$$(X_{(1)}, ..., X_{(5)})^T = (25, 32, 38, 41, 63)^T,$$

$$(R_1, ..., R_5)^T = (5, 2, 4, 1, 3)^T$$

Ranks take the values $1, 2. ..., n$ and the sum of ranks is constant

$$R_1 + \cdots + R_n = 1 + 2 + \cdots + n = \frac{n(n+1)}{2}.$$

Let us find the probability distribution of the vector of ranks

$$(R_{i_1}, \ldots, R_{i_k})^T, \quad 1 \leq i_1 < \cdots < i_k \leq n$$

Since the random variables $X_1, \ldots, X_n$ are independent and identically distributed, this vector of ranks takes $n(n - 1) \ldots (n - k + 1) = n!/(n - k)!$ different values with equal probabilities, so

$$\mathbf{P}\{(R_{i_1}, ..., R_{i_k}) = (j_1, ..., j_k)\} = \frac{(n - k)!}{n!} \qquad [4.1]$$

for all $(j_1, ..., j_k)$ made from $k$ different elements of the set $\{1, \ldots, n\}$.

In particular cases

$$\mathbf{P}\{R_i = j\} = \frac{1}{n}, \quad \mathbf{P}\{(R_{i_1}, R_{i_2}) = (j_1, j_2)\} = \frac{1}{n(n-1)}$$

$$\mathbf{P}\{(R_1, ..., R_n) = (j_1, ..., j_n)\} = \frac{1}{n!} \qquad [4.2]$$

**Theorem 4.1.** *If the distribution of the random variable $X$ is absolutely continuous then*

$$\mathbf{E}R_i = \frac{n+1}{2}, \quad \mathbf{Var}(R_i) = \frac{n^2-1}{12},$$

$$\mathbf{Cov}(R_i, R_j) = -\frac{n+1}{12}, \quad i \neq j \qquad [4.3]$$

**Proof.** We obtain

$$\mathbf{E}R_i = \sum_{j=1}^{n} j\,\mathbf{P}\{R_i = j\} = \frac{1}{n}(1 + ... + n) = \frac{n+1}{2}$$

$$\mathbf{Var}(R_i) = \mathbf{E}(R_i^2) - (\mathbf{E}R_i)^2 =$$

$$\frac{1^2 + ... + n^2}{n} - \frac{(n+1)^2}{4} = \frac{(n+1)(2n+1)}{6} - \frac{(n+1)^2}{4} = \frac{n^2-1}{12}$$

If $i \neq j$ then

$$\mathbf{Cov}(R_i, R_j) = \mathbf{E}R_i R_j - \mathbf{E}R_i \mathbf{E}R_j =$$

$$\sum\sum_{k \neq l} kl \frac{1}{n(n-1)} - \frac{(n+1)^2}{4} =$$

$$[(\sum_{k=1}^{n} k)^2 - \sum_{k=1}^{n} k^2]\frac{1}{n(n-1)} - \frac{(n+1)^2}{4} =$$

$$\frac{n(n+1)^2}{4(n-1)} - \frac{(n+1)(2n+1)}{6(n-1)} - \frac{(n+1)^2}{4} =$$

$$\frac{(n+1)}{2(n-1)}\left[\frac{n(n+1)}{2} - \frac{2n+1}{3} - \frac{n^2-1}{2}\right] =$$

$$\frac{(n+1)}{2(n-1)} \cdot \frac{-n+1}{6} = -\frac{n+1}{12}$$

$\triangle$

Let us generalize the notion of rank to the case where the distribution of the random variable $X_1$ is not necessarily absolutely continuous. If there is a group of $t$ coinciding elements

$$X_{(j-1)} < X_{(j)} = X_{(j+1)} = \cdots = X_{(j+t-1)} < X_{(j+t)} \qquad [4.4]$$

in the ordered sample $X_{(1)} \leq \cdots \leq X_{(n)}$ and $X_i = X_{(j)}$ then the rank of the sample element $X_i$ is defined as the arithmetic mean of the positions of coinciding observations

$$R_i = \frac{j + (j+1) + \cdots + (j+t-1)}{t} = j + \frac{t-1}{2} \qquad [4.5]$$

For example, if 6, 7, 5, 10, 7, 6, 7 is a sample realization then the realization of the ordered sample is 5, 6, 6, 7, 7, 7, 10, and the ranks are

$$R_1 = \frac{2+3}{2} = 2.5; \ R_2 = \frac{4+5+6}{3} = 5; \ R_3 = 1; \ R_4 = 7;$$

$$R_5 = 5; \ R_6 = 2.5; \ R_7 = 5$$

Suppose that $X_{i_1} = \cdots = X_{i_t} = X_{(j)}$; here $j$ is the number for which [4.4] is satisfied. Then sum of ranks

$$R_{i_1} + \cdots + R_{i_t} = j + (j+1) + \cdots + (j+t-1)$$

is the same as in the absolutely continuous case when the order statistics $X_{(j)}, X_{(j+1)}, \ldots, X_{(j+t-1)}$ take different values. So the sum of all ranks is the same in both cases

$$R_1 + \cdots + R_n = n(n+1)/2$$

Denote by $k$ the number of random groups with coinciding members in the ordered sample; $t_l$ is the number of the members in the $l$-th group and

$$T = \sum_{l=1}^{k} t_l(t_l^2 - 1)$$

If $t_1 = \cdots = t_n = 1$ then $T = 0$.

**Theorem 4.2.** *The means, the variances and the covariances of the ranks are*

$$\mathbf{E}R_i = \frac{n+1}{2}, \quad \mathbf{Var}(R_i) = \frac{n^2 - 1}{12} - \frac{\mathbf{E}T}{12n}$$

$$\mathbf{Cov}(R_i, R_j) = -\frac{n+1}{12} + \frac{\mathbf{E}T}{12n(n-1)} \quad (i \neq j) \qquad [4.6]$$

**Proof.** The random variables $X_i$ are independent, and identically distributed, so the random variables $R_1, \ldots, R_n$ are identically distributed and

$$\mathbf{E}R_i = \frac{1}{n}\mathbf{E}(\sum_{j=1}^{n} R_j) = \frac{n+1}{2}, \quad i = 1, \ldots, n$$

Using [4.5], write the sum of rank squares

$$\sum_{i=1}^{n} R_i^2 = \sum_{l=1}^{k} t_l(j_l + \frac{t_l - 1}{2})^2 = \sum_{l=1}^{k} t_l[j_l^2 + j_l(t_l - 1) + (t_l - 1)^2/4]$$

Denote by $j_l$ the position of the first member of the $l$-th group in the ordered sample. We obtain

$$1^2 + \cdots + n^2 = \sum_{l=1}^{k}[j_l^2 + \cdots + (j_l + t_l - 1)^2] = \sum_{l=1}^{k}[t_l j_l^2 + 2j_l(1 + \cdots + (t_l - 1)) +$$

$$(1^2 + \cdots + (t_l - 1)^2)] = \sum_{l=1}^{k} t_l[(j_l^2 + j_l(t_l - 1) + \frac{(t_l - 1)(2t_l - 1)}{6}] =$$

$$\sum_{i=1}^{n} R_i^2 + \frac{T}{12}$$

So $\sum_{i=1}^{n} R_i^2$ and using the equality

$$1^2 + \cdots + n^2 = n(n+1)(2n+1)/6$$

we obtain

$$\sum_{i=1}^{n} R_i^2 = \frac{n(n+1)(2n+1)}{6} - \frac{T}{12}$$

The random variables $R_i$ are identically distributed, so

$$\mathbf{Var}(R_i) = \mathbf{E}(R_i^2) - (\mathbf{E}R_i)^2 = \frac{(n+1)(2n+1)}{6} - \frac{\mathbf{E}T}{12n} - \frac{(n+1)^2}{4} =$$

$$\frac{n^2-1}{12} - \frac{\mathbf{E}T}{12n}$$

Note that

$$\sum_{j=1}^{n}\sum_{l=1}^{n} R_j R_l = (\frac{n(n+1)}{2})^2$$

so

$$\mathbf{E}\sum\sum_{j\neq l} R_j R_l = (\frac{n(n+1)}{2})^2 - \sum_{j=1}^{n} \mathbf{E}R_j^2 = (\frac{n(n+1)}{2})^2 -$$

$$\frac{n(n+1)(2n+1)}{6} + \frac{\mathbf{E}T}{12} =$$

$$\frac{n(n^2-1)(3n+2)}{12} + \frac{\mathbf{E}T}{12}, \quad \mathbf{E}R_j R_l = \frac{(n+1)(3n+2)}{12} + \frac{\mathbf{E}T}{12n(n-1)}$$

$$\mathbf{Cov}(R_j, R_l) = \frac{(n+1)(3n+2)}{12} + \frac{\mathbf{E}T}{12n(n-1)} - \frac{(n+1)^2}{4} =$$

$$-\frac{n+1}{12} + \frac{\mathbf{E}T}{12n(n-1)}$$

$\triangle$

**Comment 4.1.** If the sample $(X_1, \ldots, X_n)^T$ is a random vector and its distribution is symmetric then the random vectors $(X_{i_1}, \ldots, X_{i_n})^T$ have the same distribution for all permutations $(i_1, \ldots, i_n)$ of $(1, \ldots, n)$, so the distribution of ranks of such a sample is the same as in the case of a simple sample.

## 4.3.  Rank tests for independence

### 4.3.1.  *Spearman's independence test*

Suppose that

$$(X_1, Y_1)^T, \ldots, (X_n, Y_n)^T$$

is a simple sample of the random vector $(X, Y)^T$ with the cumulative distribution function $F = F(x, y) \in \mathcal{F}$; here $\mathcal{F}$ is a non-parametric class of absolutely continuous two-dimensional distribution functions.

**Independence hypothesis:** (of two random variables)

$$H_0 : F \in \mathcal{F}_0$$

where $\mathcal{F}_0 \subset \mathcal{F}$ is the set of two-dimensional cumulative distribution functions which are a product of marginal distribution functions:

$$F(x, y) = F_1(x) F_2(y) \quad \text{for all} \quad (x, y) \in \mathbf{R}^2$$

where $F_1(x) = \mathbf{P}\{X \leq x\}$ and $F_2(y) = \mathbf{P}\{Y \leq y\}$.

Denote by

$$R_{11}, \ldots, R_{1n} \quad \text{and} \quad R_{21}, \ldots, R_{2n}$$

the ranks of the elements of the samples

$$X_1, \ldots, X_n \quad \text{and} \quad Y_1, \ldots, Y_n$$

respectively. These ranks are used for independence hypothesis test construction.

**Definition 4.2.** The empirical correlation coefficient between the ranks of the first and the second samples

$$r_S = \frac{\sum_{j=1}^{n}(R_{1j} - \bar{R}_1)(R_{2j} - \bar{R}_2)}{[\sum_{j=1}^{n}(R_{1j} - \bar{R}_1)^2 \sum_{j=1}^{n}(R_{2j} - \bar{R}_2)^2]^{1/2}}$$

$$\bar{R}_i = \frac{1}{n}\sum_{j=1}^{n} R_{ij} = \frac{n+1}{2} \qquad [4.7]$$

is called Spearman's rank correlation coefficient.

Arrange the observations $(X_i, Y_i)^T$, $i = 1, 2, ..., n$, so that $Y_i$ are written in increasing order and the obtained vectors of $X$-s and $Y$-s are replaced by their ranks. Instead of the line $(R_{21}, ..., R_{2n})$, we obtain the line $(1, 2, ..., n)$, and instead of $(R_{11}, ..., R_{1n})$, we obtain the line with elements denoted by $R_1, ..., R_n$. The value of the coefficient $r_S$ does not change because the same pairs of ranks are used. We have:

$$r_S = \frac{\sum_{i=1}^{n}(R_i - \frac{n+1}{2})(i - \frac{n+1}{2})}{[\sum_{i=1}^{n}(R_i - \frac{n+1}{2})^2 \sum_{i=1}^{n}(i - \frac{n+1}{2})^2]^{1/2}}$$

Since

$$\sum_{i=1}^{n}\left(R_i - \frac{n+1}{2}\right)^2 = \sum_{i=1}^{n}\left(i - \frac{n+1}{2}\right)^2 = n\mathbf{Var}(R_i) = \frac{n(n^2-1)}{12}$$

$$\sum_{i=1}^{n}\left(R_i - \frac{n+1}{2}\right)\left(i - \frac{n+1}{2}\right) =$$

$$\sum_{i=1}^{n} iR_i - \frac{n(n+1)^2}{4} = \frac{n(n^2-1)}{12} - \frac{1}{2}\sum_{i=1}^{n}(R_i - i)^2$$

the Spearman's rank correlation coefficient can be written in the following form.

**Spearman's rank correlation coefficient:**

$$r_S = \frac{\frac{1}{n}\sum_{i=1}^{n} iR_i - \left(\frac{n+1}{2}\right)^2}{(n^2-1)/12} = 1 - \frac{6}{n(n^2-1)}\sum_{i=1}^{n}(R_i - i)^2 \quad [4.8]$$

By the independence hypothesis, the distribution of the random vector $(R_1, \ldots, R_n)^T$ coincides with the distribution of the random vector $(R_{11}, \ldots, R_{1n})^T$, so it does not depend on unknown parameters and depends only on sample size $n$.

**Spearman's independence test:** independence hypothesis $H_0$ is rejected with a significance level not greater than $\alpha$ if

$$r_s \leq c_1 \quad \text{or} \quad r_s \geq c_2 \quad\quad\quad [4.9]$$

where $c_1$ is the maximum and $c_2$ is the minimum possible value of $r_S$ satisfying inequalities

$$\mathbf{P}\{r_s \leq c_1\} \leq \alpha/2, \quad \mathbf{P}\{r_s \geq c_2\} \leq \alpha/2$$

For small $n$, the probability distribution of the random variable $r_S$, and hence the critical values, can be found using the probability distribution of the random vector $(R_1, \ldots, R_n)^T$, given in [4.2] because $r_S$ is a function of this vector.

If $n$ is large then the central limit theorem is used to find the asymptotic distribution of $r_S$. Let us find the first two moments of $r_s$. Under the independence hypothesis

$$\mathbf{E}\sum_i i\left(R_i - \frac{n+1}{2}\right) = 0, \quad \mathbf{Var}\left(\sum_i iR_i\right) = \mathbf{Var}(R_1)\sum_i i^2 +$$

$$\mathbf{Cov}(R_1, R_2)\sum_{i \neq j} ij =$$

$$\frac{n^2-1}{12}\frac{n(n+1)(2n+1)}{6} - \frac{n+1}{12}\left[\left(\frac{n(n+1)}{2}\right)^2 - \frac{n(n+1)(2n+1)}{6}\right]$$

$$= \frac{n^2(n+1)^2(n-1)}{144}$$

so, using [4.8] we obtain

$$\mathbf{E}r_S = 0, \quad \mathbf{Var}(r_S) = \frac{1}{n-1}$$

If $n$ is large then the distribution of the random variable $r_S$ is approximated by the normal distribution

$$Z_n = \sqrt{n-1}\, r_S \xrightarrow{d} Z \sim N(0,\, 1) \qquad [4.10]$$

Using the normal approximation we obtain the following test.

**Asymptotic Spearman's independence test:** the independence hypothesis is rejected with an asymptotic significance level $\alpha$ if

$$|Z_n| > z_{\alpha/2} \qquad [4.11]$$

For medium-sized samples the probability distribution of the statistic

$$t_n = \sqrt{n-2}\, \frac{r_S}{\sqrt{1 - r_S^2}}$$

is approximated by Student's distribution $S(n-2)$.

**Spearman's independence test based on Student's approximation:** the independence hypothesis is rejected with an approximate significance level $\alpha$ if

$$|t_n| > t_{\alpha/2}(n-2) \qquad [4.12]$$

**Modification when *ex aequo* exist.** Suppose that some ranks coincide. Formula [4.7] implies that Spearman's rank correlation coefficient can be computed using the following formula

$$r_S = \frac{\sum_{j=1}^{n}(R_{1j} - \frac{n+1}{2})(R_{2j} - \frac{n+1}{2})}{[\sum_{j=1}^{n}(R_{1j} - \frac{n+1}{2})^2 \sum_{j=1}^{n}(R_{2j} - \frac{n+1}{2})^2]^{1/2}} =$$

$$\frac{\sum_{j=1}^{n} R_{1j}R_{2j} - n(\frac{n+1}{2})^2}{[(\sum_{j=1}^{n} R_{1j}^2 - n(\frac{n+1}{2})^2)(\sum_{j=1}^{n} R_{2j}^2 - n(\frac{n+1}{2}))^2]^{1/2})}$$

Using [4.6] we see that *under the independence hypothesis* the mean of the numerator is zero and the variance is

$$= n(\frac{n^2 - 1}{12})^2(1 - \frac{ET_X}{n^3 - n})(1 - \frac{ET_Y}{n^3 - n})(1 + \frac{1}{n - 1})$$

where

$$T_X = \sum_{l=1}^{k} t_l(t_l^2 - 1)$$

$k$ is the number of random groups with equal members in the ordered sample of $x$-s and $t_l$ is the number of members in the $l$-th group. $T_Y$ is defined analogously.

It follows from the proof of Theorem 4.2 that

$$\sum_{j=1}^{n} R_{1j}^2 - n\left(\frac{n+1}{2}\right)^2 = \frac{n(n^2 - 1)}{12} - \frac{T_X}{12}$$

and the expression under the square-root sign in the denominator has the form

$$\left(\frac{n(n^2 - 1)}{12}\right)^2 (1 - \frac{T_X}{n^3 - n})(1 - \frac{T_Y}{n^3 - n})$$

So, if some ranks coincide and $n$ is large, the mean of the random variable $r_S$ is close to $0$ and the variance is close to $1/n$, so the same approximation, [4.10], is used and the asymptotic tests have the form [4.11] or [4.12].

**Example 4.1.** The results $X_i$ and $Y_i$ of mathematics and foreign language tests of $n = 50$ pupils are given in the following table.

| $i$ | 1 | 2 | 3 | 4 | 5 | 6 | 7 | 8 | 9 | 10 | 11 | 12 | 13 |
|---|---|---|---|---|---|---|---|---|---|---|---|---|---|
| $X_i$ | 59 | 63 | 72 | 55 | 50 | 46 | 67 | 61 | 67 | 53 | 39 | 41 | 62 |
| $Y_i$ | 50 | 55 | 53 | 54 | 59 | 52 | 57 | 58 | 57 | 60 | 49 | 59 | 59 |

| $i$ | 14 | 15 | 16 | 17 | 18 | 19 | 20 | 21 | 22 | 23 | 24 | 25 | 26 |
|---|---|---|---|---|---|---|---|---|---|---|---|---|---|
| $X_i$ | 51 | 64 | 52 | 54 | 59 | 64 | 32 | 48 | 65 | 62 | 53 | 65 | 58 |
| $Y_i$ | 50 | 66 | 5! | 59 | 60 | 58 | 57 | 52 | 57 | 52 | 58 | 58 | 64 |

| $i$ | 27 | 28 | 29 | 30 | 31 | 32 | 33 | 34 | 35 | 36 | 37 | 38 | 39 |
|---|---|---|---|---|---|---|---|---|---|---|---|---|---|
| $X_i$ | 51 | 53 | 64 | 64 | 61 | 65 | 40 | 52 | 38 | 56 | 49 | 60 | 52 |
| $Y_i$ | 55 | 54 | 56 | 57 | 59 | 62 | 54 | 55 | 51 | 64 | 55 | 50 | 50 |

| $i$ | | 4041 | 42 | 43 | 44 | 45 | 46 | 47 | 48 | 49 | 50 |
|---|---|---|---|---|---|---|---|---|---|---|---|
| $X_i$ | 65 | 68 | 58 | 47 | 39 | 59 | 60 | 42 | 51 | 52 | 65 |
| $Y_i$ | 54 | 59 | 59 | 57 | 42 | 49 | 50 | 37 | 46 | 48 | 60 |

Are the two test results realizations of two independent random variables?

Arrange the $x$ values in increasing order of the results of the mathematics test and find the ranks $R_{1i}$. Arrange the $y$ values in increasing order of the results of the foreign language test and find the ranks $R_{2i}$. Since there are coinciding results in both tests, the ranks are found using [4.5]. The values of the ranks are given in the following table.

| $i$ | 1 | 2 | 3 | 4 | 5 | 6 | 7 | 8 | 9 | 10 |
|---|---|---|---|---|---|---|---|---|---|---|
| $R_{1i}$ | 29 | 37 | 50 | 24 | 12 | 8 | 47.5 | 33.5 | 47.5 | 21 |
| $R_{2i}$ | 9 | 23.5 | 17 | 19.5 | 40 | 15 | 29.5 | 34.5 | 29.5 | 45 |

| $i$ | 11 | 12 | 13 | 14 | 15 | 16 | 17 | 18 | 19 | 20 |
|---|---|---|---|---|---|---|---|---|---|---|
| $R_{1i}$ | 3.5 | 6 | 35.5 | 14 | 39.5 | 17.5 | 23 | 29 | 39.5 | 1 |
| $R_{2i}$ | 5.5 | 40 | 40 | 9 | 50 | 12.5 | 40 | 45 | 34.5 | 29.5 |

| $i$ | 21 | 22 | 23 | 24 | 25 | 26 | 27 | 28 | 29 | 30 |
|---|---|---|---|---|---|---|---|---|---|---|
| $R_{1i}$ | 10 | 44 | 35.5 | 21 | 44 | 26.5 | 14 | 21 | 39.5 | 39.5 |
| $R_{2i}$ | 15 | 29.5 | 15 | 34.5 | 34.5 | 48.5 | 23.5 | 19.5 | 26 | 29.5 |

| $i$ | 31 | 32 | 33 | 34 | 35 | 36 | 37 | 38 | 39 | 40 |
|---|---|---|---|---|---|---|---|---|---|---|
| $R_{1i}$ | 33.5 | 44 | 5 | 17.5 | 2 | 25 | 11 | 31.5 | 17.5 | 44 |
| $R_{2i}$ | 40 | 47 | 19.5 | 23.5 | 12.5 | 48.5 | 23.5 | 9 | 9 | 19.5 |

| $i$ | 41 | 42 | 43 | 44 | 45 | 46 | 47 | 48 | 49 | 50 |
|---|---|---|---|---|---|---|---|---|---|---|
| $R_{1i}$ | 49 | 26.5 | 9 | 3.5 | 29 | 31.5 | 7 | 14 | 17.5 | 44 |
| $R_{2i}$ | 40 | 40 | 29.5 | 2 | 5.5 | 9 | 1 | 3 | 4 | 45 |

Spearman's rank correlation coefficient is

$$r_S = \frac{\sum_{j=1}^{50} R_{1j}R_{2j} - 50(\frac{51}{2})^2}{[(\sum_{j=1}^{50} R_{1j}^2 - 50(\frac{51}{2})^2)(\sum_{j=1}^{50} R_{1j}^2 - 50(\frac{51}{2})^2)]^{1/2}} = 0.39003$$

Using the asymptotic Spearman's test, based on the normal approximation, we obtain

$$Z_n = \sqrt{n-1}\,r_S = \sqrt{49}\,0.39003 = 2.73024$$

The asymptotic $P$-value is

$$pv_a = 2(1 - \Phi(2.73024)) = 0.00633$$

Using the asymptotic Spearman's test based on Student's approximation we obtain

$$t_n = \sqrt{50-2}\frac{0.39003}{\sqrt{1-0.39003^2}} = 0.00511$$

The hypothesis is rejected by both tests because in both cases the $P$-value is small. So we conclude that the results of the two tests are dependent. Since $r_S > 0$ pupils with better results in the mathematics test have a tendency to obtain better results in the foreign language test.

### 4.3.2. Kendall's independence test

**Definition 4.3.** If a number is written before a smaller number then we say that there is an *inversion*. For example, in the permutation $(5, 3, 1, 4, 2)$ there are $4+2+1 = 7$ inversions.

The number $I_n$ of inversions in the permutation $(R_1, ..., R_n)$ takes values from 0 (the numbers are arranged in increasing order) to $N_n = n(n-1)/2$ (the numbers are arranged in decreasing order).

**Theorem 4.3.** *Under the independence hypothesis, the characteristic function of the number of inversions $I_n$ is*

$$\varphi_n(t) = \mathbf{E}e^{itI_n} = \frac{1}{n!}\prod_{j=1}^{n}\frac{e^{itj}-1}{e^{it}-1} = \prod_{j=1}^{n}\left(\frac{1+e^{it}+\cdots+e^{it(j-1)}}{j}\right)$$

[4.13]

**Proof.** We obtain

$$\varphi_n(t) = \mathbf{E}e^{itI_n} = \sum_{k=0}^{N_n}e^{itk}\frac{\nu_n(k)}{n!}$$

[4.14]

where $\nu_n(k)$ is the number of permutations $\{i_1, ..., i_n\}$ of $\{1, 2...., n\}$ having $k$ inversions.

Let us express the number $\nu_n(k)$ by numbers $\nu_{n-1}(l)$, $k-(n-1) \le l \le \min(k, \ n-1)$.

Note that $n!$ permutations of $\{1, 2...., n\}$ can be obtained from $(n-1)!$ permutations of $\{1, 2...., n-1\}$ putting "$n$" in all possible places.

For example, from 2! permutations $(1, 2)$ and $(2, 1)$ we obtain 3! permutations putting 3 in all possible places: $(1, 2, 3)$, $(1, 3, 2)$, $(3, 1, 2)$, $(2, 1, 3)$, $(2, 3, 1)$ and $(3, 2, 1)$.

The permutation of $(1, 2, \ldots, n)$ having $k$ inversions is obtained from the permutations of $(1, 2, \ldots, n-1)$ having $k-l$ inversions putting $n$ in the position such that after this position there are $l$ numbers ($l = 0, \ldots, n-1$, if $k \geq n-1$; $l = 0, \ldots, k$, if $0 \leq k < n-1$). So the equalities

$$\nu_n(k) = \nu_{n-1}(k) + \nu_{n-1}(k-1) + \cdots + \nu_{n-1}(k-(n-1)), \quad k \geq n-1$$

$$\nu_n(k) = \nu_{n-1}(k) + \nu_{n-1}(k-1) + \cdots + \nu_{n-1}(0), \quad 0 \leq k < n-1$$

$$\text{[4.15]}$$

hold.

For any $l = 0, 1, \ldots, N_{n-1}$, the term $\nu_{n-1}(l)$ is included in expression [4.15] of $\nu_n(k)$ if $k = l, \ldots, l+n-1$, because the number of inversions can increase by $0, 1, \ldots, n-1$. Hence

$$\varphi_n(t) = \sum_{k=0}^{N_n} e^{itk} \frac{\nu_n(k)}{n!} = \sum_{l=0}^{N_{n-1}} \frac{\nu_{n-1}(l)}{n!} \sum_{k=l}^{l+n-1} e^{itk} =$$

$$\frac{\varphi_{n-1}(t)}{n} [1 + e^{it} + \cdots + e^{it(n-1)}] = \frac{e^{itn} - 1}{n(e^{it} - 1)} \varphi_{n-1}(t)$$

Using this equality and taking into account that

$$\varphi_1(t) = \nu_1(0) = 1, \quad \varphi_2(t) = \frac{\nu_2(0)}{2} + \frac{\nu_2(1)}{2} e^{it} = \frac{e^{2it} - 1}{2(e^{it} - 1)}$$

we obtain [4.13].

$\triangle$

**Definition 4.4.** The random variable

$$r_K = 1 - \frac{4I_n}{n(n-1)}, \quad -1 \leq r_K \leq 1 \qquad \text{[4.16]}$$

is called *Kendall's rank correlation coefficient.*

**Theorem 4.4.** *Under the independence hypothesis, the first two moments of $r_K$ are*

$$\mathbf{E}r_K = 0, \quad \mathbf{Var}(r_K) = \frac{2(2n+5)}{9n(n-1)}$$

*and the convergence*

$$\frac{r_K}{\sqrt{\mathbf{Var}(r_K)}} \overset{d}{\to} Z \sim N(0,1) \quad as \quad n \to \infty \qquad [4.17]$$

*holds.*

**Proof.** From the form [4.13] of the characteristic functions we see that $I_n$ is a sum of independent random variables

$$I_n = U_1 + U_2 + \cdots + U_n \qquad [4.18]$$

When $U_1 = 0$, the random variable $U_j$ takes values $0, 1, ..., j - 1$ with equal probabilities $1/j$, $j = 2...., n$, because the characteristic function of $U_j$ is

$$\psi(t) = \mathbf{E}e^{itU_j} = \sum_{k=0}^{j-1} e^{itk} \frac{1}{j}$$

So

$$\mathbf{E}U_j = \frac{1}{j} \sum_{i=1}^{j-1} i = \frac{1}{j} \frac{j(j-1)}{2} = \frac{j-1}{2}$$

$$\mathbf{E}U_j^2 = \frac{(j-1)(2j-1)}{6}, \quad \mathbf{Var}U_j = \frac{j^2-1}{12}$$

hence

$$\mathbf{E}I_n = \sum_{j=1}^{n} \frac{j-1}{2} = \frac{n(n-1)}{4}, \quad \mathbf{E}R_K = 0$$

$$\mathbf{Var}(I_n) = \sum_{j-1}^{n} \frac{j^2-1}{12} = \frac{n(n-1)(2n+5)}{72}$$

$$\mathbf{Var}(r_K) = \frac{16\mathbf{Var}(I_n)}{n^2(n-1)^2} = \frac{2(2n+5)}{9n(n-1)}$$

Let us prove that the central limit theorem can be applied. It is sufficient to verify Lindeberg's condition for the number of inversions $I_n$

$$\frac{1}{\mathbf{Var}(I_n)} \sum_{j=1}^{n} \int_{\frac{|x-\mathbf{E}U_j|}{\sqrt{\mathbf{Var}(I_n)}}>\epsilon} (x-\mathbf{E}U_j)^2 dF_{U_j}(x) \to 0, \quad \text{as} \quad n \to \infty$$

The random variables $U_j$ take values from 1 to $j-1$, $j = 1, 2, ..., n$, $\sqrt{\mathbf{Var}(I_n)} = O(n^{3/2})$, so

$$\frac{|x-\mathbf{E}U_j|}{\sqrt{\mathbf{Var}(I_n)}} \le \frac{n-1}{O(n^{3/2})} < \varepsilon$$

if $n$ is sufficiently large and all integration regions are empty. So Lindeberg's condition is verified and the central limit theorem can be applied to the sum $I_n$, and hence can be applied to the coefficient $r_K$, which is a linear function of $I_n$.

$\triangle$

For small $n$, tables of critical values of the statistic $r_K$ are available in most statistical packages.

**Kendall's independence test:** the independence hypothesis is rejected with a significance level no greater than $\alpha$ if

$$r_K \le c_1 \quad \text{or} \quad r_K \ge c_2 \qquad [4.19]$$

where $c_1$ is the maximum and $c_2$ is the minimum possible value of $r_K$, verifying the inequalities

$$\mathbf{P}\{r_K \le c_1\} \le \alpha/2. \quad \mathbf{P}\{r_K \ge c_2\} \le \alpha/2$$

If $n$ is large then the normal approximation is used.

**Asymptotic    Kendall's    independence    test:    the** independence hypothesis is rejected with an asymptotic significance level $\alpha$ if

$$\left| \frac{r_K}{\sqrt{\mathbf{Var}(r_K)}} \right| > z_{\alpha/2} \qquad [4.20]$$

A more general definition of Kendall's rank correlate coefficient is used when coinciding values are possible in both samples.

Suppose that the pairs $(X_i, Y_i)^T$ and $(X_j, Y_j)^T$ are *concordant* if the signs of $X_j - X_i$ and $Y_j - Y_i$ coincide: $(X_j - X_i)(Y_j - Y_i) > 0$. The pairs are *discordant* if $(X_j - X_i)(Y_j - Y_i) < 0$. Set

$$A_{ij} = \left\{ \begin{array}{l} 1, \text{ if } (X_j - X_i)(Y_j - Y_i) > 0, \\ -1, \text{ if } X_j - X_i)(Y_j - Y_i) < 0, \\ 0, \text{ if } (X_j - X_i)(Y_j - Y_i) = 0 \end{array} \right.$$

**Definition 4.5.** The statistic

$$\tau_a = \frac{2}{n(n-1)} \sum_{i<j} A_{ij} \qquad [4.21]$$

is called *Kendall's tau$_a$ correlation coefficient.*

$\sum_{i<j} A_{ij}$ is *the difference between the numbers of concordant and discordant pairs.*

**Comment 4.2.** If there is no coinciding value in both samples then $r_K = \tau_a$.

Indeed, if we set

$$h_{ij} = \left\{ \begin{array}{l} 1, \text{ if } R_i > R_j, \\ 0, \text{ if } R_i < R_j \end{array} \right.$$

then the number of inversions can be written as the sum

$$I_n = \sum_{i<j} h_{ij} \qquad [4.22]$$

Since the number $-1$ is found $\sum_{i<j} h_{ij}$ times and the number 1 is found $n(n-1)/2 - \sum_{i<j} h_{ij}$ times in the sum $\sum_{i<j} A_{ij}$, we have

$$\sum_{i<j} A_{ij} = n(n-1)/2 - 2\sum_{i<j} h_{ij} = n(n-1)/2 - 2I_n$$

hence

$$\frac{2}{n(n-1)} \sum_{i<j} A_{ij} = 1 - \frac{4I_n}{n(n-1)}$$

**Comment 4.3.** If coinciding values are present then Kendall's $tau_a$ correlation coefficient cannot be equal to 1 even if $X_i = Y_i$ for all $i$ because not all terms are equal to 1 in the sum $\sum_{i<j} A_{ij}$ and this sum has $n(n-1)/2$ terms.

Let us consider the following modification.

The random variables $A_{ij}$ are written in the form $A_{ij} = U_{ij}V_{ij}$; here

$$U_{ij} = \begin{cases} 1, & \text{if } X_j - X_i > 0, \\ -1, & \text{if } X_j - X_i < 0 \\ 0, & \text{if } X_j - X_i = 0. \end{cases} , \quad V_{ij} = \begin{cases} 1, & \text{if } Y_j - Y_i > 0, \\ -1, & \text{if } Y_j - Y_i < 0 \\ 0, & \text{ifi } Y_j - Y_i = 0 \end{cases}$$

**Definition 4.6** The statistic

$$\tau_b = \frac{\sum_{i=1}^n \sum_{j=1}^n U_{ij}V_{ij}}{[(\sum_{i=1}^n \sum_{j=1}^n U_{ij}^2)(\sum_{i=1}^n \sum_{j=1}^n V_{ij}^2)]^{1/2}} \qquad [4.23]$$

is called *Kendall's tau$_b$ correlation coefficient.*

**Comment 4.4.** If in both samples there is no coinciding value then

$$\sum_{i=1}^{n}\sum_{j=1}^{n}U_{ij}^2 = \sum_{i=1}^{n}\sum_{j=1}^{n}V_{ij}^2 = n(n-1)$$

so $\tau_b = \tau_a = r_K$.

**Comment 4.5.** If the samples have $k_X$ and $k_Y$ groups of coinciding elements and the numbers of elements in the $s$-th and the $r$-th groups are $u_s$ and $v_r$, respectively, Kendall's $tau_b$ correlation coefficient can be written in the form

$$\tau_b = \frac{\sum_{i=1}^{n}\sum_{j=1}^{n}U_{ij}V_{ij}}{[(n(n-1) - \sum_{s=1}^{k_X}u_s(u_s-1))(n(n-1) - \sum_{r=1}^{k_Y}v_r(v_r-1))]^{1/2}}$$

[4.24]

because there are $u_s(u_s-1)$ pairs with $U_{ij} = 0$ in the $s$-th group, so

$$\sum_{i=1}^{n}\sum_{j=1}^{n}U_{ij}^2 = n(n-1) - \sum_{s=1}^{k_X}u_s(u_s-1)$$

The analogous equality holds for the second sample.

So, if coinciding values exist then $|\tau_b| > |\tau_a|$ because the numerators coincide and the denominator in the expression $\tau_b$ is smaller. If $X_i = Y_i$ for all $i$ then $U_{ij} = V_{ij}$, hence $\tau_b = 1$.

Testing the independence hypothesis, the normal approximation $N(0, V_S)$ of the numerator

$$S = \sum_{i<j}U_{ij}V_{ij}$$

is used; here

$$V_S = \frac{\nu_0 - \nu_u - \nu_v}{18} + \frac{\nu_{uv1}}{2n(n-1)} + \frac{\nu_{uv2}}{9n(n-1)(n-2)}$$

$$\nu_0 = n(n-1)(2n+5), \quad \nu_u = \sum_{s=1}^{k_X} u_s(u_s-1)(2u_s+5)$$

$$\nu_v = \sum_{r=1}^{k_Y} v_r(v_r-1)(2v_r+5)$$

$$\nu_{uv1} = \sum_{s=1}^{k_X} u_s(u_s-1) \sum_{r=1}^{k_Y} v_r(v_r-1)$$

$$\nu_{uv2} = \sum_{r=1}^{k_X} u_s(u_s-1)(u_s-2) \sum_{r=1}^{k_Y} v_r(v_r-1)(v_r-2)$$

**Asymptotic Kendall's independence test:** the independence hypothesis is rejected with an asymptotic significance level $\alpha$ if

$$\left| \frac{S}{\sqrt{V_{S)}}} \right| > z_{\alpha/2} \qquad\qquad [4.25]$$

**Comment 4.6.** The correlation of the random variables is also estimated by the *Goodman–Kruskal gamma coefficient*:

$$\gamma = \frac{\sum\limits_{A_{ij}=1} A_{ij} - \sum\limits_{A_{ij}=-1} A_{ij}}{\sum\limits_{A_{ij}\neq 0} A_{ij}}.$$

The coefficient is equal to the difference between the numbers of concordant and discordant pairs divided by the number of non-coinciding pairs. This coefficient also coincides with Kendall's correlation coefficient $r_K$ if there is no coinciding observation.

**Example 4.2.** (continuation of Example 4.1) Using the data in Example 4.1 and the asymptotic Kendall's independence test, verify the hypothesis formulated in Example 4.1.

Summing the products $U_{ij}V_{ij}$ we obtain

$$\sum_{i=1}^{n}\sum_{j=1}^{n}U_{ij}V_{ij} = 2S = 630$$

In the first sample there are $k_X = 12$ groups with coinciding values: 6 have 2, 3 have 3, 2 have 4 and 1 have 5 elements, respectively. In the second sample there are $k_Y = 11$ groups with coinciding values: 3 have 2, 2 have 3, 3 have 4, 1 have 5, 1 have 6 and 1 has 7 elements. So

$$\sum_{s=1}^{k_X}u_s(u_s-1) = 6\cdot2+3\cdot6+2\cdot12+1\cdot20 = 74, \quad \sum_{r=1}^{k_Y}v_r(v_r-1) = 146$$

$$\tau_b = \frac{630}{[(50\cdot49-74)(50\cdot49-74)]^{1/2}} = 0.26926$$

$$\nu_0 = 50\cdot49\cdot105 = 257250$$

$$\nu_u = 6\cdot2\cdot1\cdot9+3\cdot3\cdot2\cdot11+2\cdot4\cdot3\cdot13+1\cdot5\cdot4\cdot15 = 918, \quad \nu_v = 2262$$

$$\nu_{uv1} = 74\cdot146 = 10804, \quad \nu_{uv2} = 125\cdot474 = 59724$$

So the estimator of the variance of the statistic $S$ is

$$V_S = \frac{257250-918-2262}{18}+\frac{10804}{2\cdot50\cdot49}+\frac{59724}{9\cdot50\cdot49\cdot48} = 14117,26$$

We obtain

$$\frac{S}{\sqrt{V_S}} = 2.65116.$$

The asymptotic $P$-value of the test [4.25] is $pv_a = 2(1-\Phi(2.65116)) = 0.00802$. As in the case of Spearman's test, Kendall's test rejects the hypothesis because the $P$-value is small.

**Comment 4.7.** Coefficients $r_S$ and $r_K$ are closely related.

Let us define a weighted sum of inversions taking into account the distance between the ranks

$$I_n^* = \frac{3}{n+1} \sum_{i<j} (j-i) h_{ij}$$

It can be shown (see exercise 4.1) that

$$I_n^* = \frac{3}{2(n+1)} \sum_i (R_i - i)^2, \quad r_S = 1 - \frac{4 I_n^*}{n(n-1)}$$

By comparing with [4.16], we see that the only difference between $r_S$ and $r_K$ is that in the definition of the first and the second coefficients the weighted sum $I_n^*$ and the non-weighted sum $I_n^*$ of the inversions are used respectively.

Pearson's correlation coefficient between $r_S$ and $r_K$ (see exercise 4.2) is

$$\rho(r_S, r_K) = \frac{2(n+1)}{\sqrt{2n(2n+5)}}$$

The coefficient decreases from 1 ($n = 2$) to 0.98 ($n = 5$), and increases again to 1 as $n \to \infty$. So both statistics are very similar and the tests based on them are practically equivalent for all finite $n$.

### 4.3.3. ARE of Kendall's independence test with respect to Pearson's independence test under normal distribution

The ARE (asymptotic relative efficiency) characterizes the behavior of the powers of tests in the neighborhood of the hypothetical value of some parameters. More precisely, the behavior of the powers of tests is investigated when a sequence of alternatives approaches the hypothetical value with a given rate as the sample sizes go to infinity.

Suppose that independent random vectors $(X_i, \ Y_i)^T$, $i = 1, ..., n$, have a two-dimensional normal distribution with correlation coefficient $\rho$.

In the case of a normal distribution, the random variables $X$ and $Y$ are independent if and only if Pearson's correlation coefficient $\rho = 0$. Pearson's independence test is based on Pearson's empirical correlation coefficient

$$\hat{\rho} = r = \frac{\sum_i (X_i - \bar{X})(Y_i - \bar{Y})}{\sqrt{\sum_i (X_i - \bar{X})^2 \sum_i (Y_i - \bar{Y})^2}} \qquad [4.26]$$

If $\rho = 0$, the pdf of $r$ is

$$f_r(x) = \frac{1}{\sqrt{\pi}} \frac{\Gamma(\frac{n-1}{2})}{\Gamma(\frac{n-2}{2})} (1 - x^2)^{\frac{n-4}{2}}, \quad -1 < x < 1$$

Define the rv

$$U = \sqrt{n-2} \frac{r}{\sqrt{1-r^2}} \qquad [4.27]$$

The pdf of $U$ is

$$g(u) = \frac{1}{\sqrt{(n-2)\pi}} \frac{\Gamma(\frac{n-1}{2})}{\Gamma(\frac{n-2}{2})} (1 + \frac{u^2}{n-2})^{-(n-1)/2}$$

i.e. the random variable $U$ has a Student's distribution with $n - 2$ degrees of freedom.

**Independence test based on Pearson's empirical correlation coefficient:** the independence hypothesis is rejected against the alternatives $H_1 : \rho > 0$, $H_2 : \rho < 0$ and $H_3 : \rho \neq 0$, respectively, with a significance level $\alpha$, if

$$U > t_\alpha(n-2), \quad U < -t_\alpha(n-2)\}, \quad |U| > t_{\alpha/2}(n-2)\}$$

respectively.

**Theorem 4.5.** *Suppose that the independent random vectors* $(X_i, Y_i)^T$, $i = 1, ..., n$, *have a two-dimensional normal distribution with correlation coefficient* $\rho$. *Then the ARE of the test based on Kendall's empirical correlation coefficient with respect to the test based on Pearson's empirical correlation coefficient* $r$ *is*

$$e(r_K, r) = \frac{9}{\pi^2} \approx 0.912$$

**Proof.** Set

$$\mu_1(\rho) = \mathbf{E}_\rho(r_K), \quad \sigma_1^2(0) = \lim_{n \to \infty} n \mathbf{Var}_0(r_K)$$

$$\mu_2(\rho) = \lim_{n \to \infty} \mathbf{E}_\rho(r), \quad \sigma_2^2(0) = \lim_{n \to \infty} n \mathbf{Var}_0(r)$$

Using formula [1.5] (where $\delta = 1/2$), we must find $\mu_1'(0)$, $\mu_2'(0)$, $\sigma_1(0)$ and $\sigma_2(0)$.

Since $r_K = 1 - 4I_n/(n(n-1))$, to find $\mu_1(\rho)$ it is sufficient to find the mean of the number of inversions $I_n$ (see Definition 4.3) under alternatives. The number of inversions $I_n$ can be written as follows

$$I_n = \sum_{i<j} \tilde{h}_{ij}, \quad \tilde{h}_{ij} = \frac{1}{2}\{1 - sign(X_i - X_j)sign(Y_i - Y_j)\}$$

Indeed

$$\tilde{h}_{ij} = 1 \iff (X_i - X_j)(Y_i - Y_j) < 0 \iff$$

$$(R_{1i} - R_{1j})(R_{2i} - R_{2j}) < 0 \iff (k - l)(R_k - R_l) < 0;$$

where $k = R_{2i}, l = R_{2j}$. The condition $(k - l)(R_k - R_l) < 0$ is equivalent to the presence of inversion.

Since there are $n(n-1)/2$ equally distributed terms in the sum $I_n = \sum_{i<j} \tilde{h}_{ij}$

$$\mathbf{E}I_n = \frac{n(n-1)}{2}\mathbf{E}(\tilde{h}_{ij})$$

$$\mu_1(\rho) = \mathbf{E}_\rho r_K = \mathbf{E}_\rho(1 - \frac{4I_n}{n(n-1)}) = 1 - 2\mathbf{E}_\rho(\tilde{h}_{ij}) =$$

$$\mathbf{E}_\rho(sign(X_i - X_j)sign(Y_i - Y_j))$$

Since the independent random vectors $(X_i, \ Y_i)^T$, $i = 1, ..., n$, are normally distributed with correlation coefficient $\rho$, the vector $(U, V)^T$ with coordinates $U = X_i - X_j$ and $V = Y_i - Y_j$ is also normally distributed with zero mean and the same correlation coefficient $\rho$. So

$$\mu_1(\rho) = \int_{-\infty}^{\infty} \int_{-\infty}^{\infty} sign(u)sign(v) \ \phi(u, \ v; \ \rho)du\,dv =$$

$$2\int_0^\infty \left[\int_0^\infty - \int_{-\infty}^0\right] \phi(u, \ v; \ \rho)dv\,du.$$

The integral does not depend on the values of the variances of the random variables $U$ and $V$ (changing the variables $x = u/\sigma_1$ and $y = v/\sigma_2$ does not change the limits of integrals), so the probability density function of the standard two-dimensional normal distribution can be considered

$$\phi(u, \ v; \ \rho) = \frac{1}{2\pi\sqrt{1-\rho^2}} \exp\{-\frac{1}{2(1-\rho^2)}[u^2 - 2\rho uv + v^2]\} =$$

$$\frac{1}{\sqrt{2\pi(1-\rho^2)}}e^{-\frac{(v-\rho u)^2}{2(1-\rho^2)}} \frac{1}{\sqrt{2\pi}}e^{-\frac{u^2}{2}}$$

We obtain

$$\mu_1(\rho) = 2\int_0^\infty \left[1 - 2\Phi\left(\frac{-\rho u}{\sqrt{1-\rho^2}}\right)\right] \frac{1}{\sqrt{2\pi}}e^{-\frac{u^2}{2}}\,du$$

$$\Phi(x) = \int_{-\infty}^x \frac{1}{\sqrt{2\pi}}e^{-\frac{y^2}{2}}\,dy$$

Differentiating with respect to $\rho$ we obtain

$$\mu_1'(\rho) = \frac{2}{\pi(1-\rho^2)^{3/2}} \int_0^\infty ue^{-\frac{u^2}{2(1-\rho^2)}}\,du = \frac{2}{\pi\sqrt{1-\rho^2}}, \quad \mu_1'(0) = \frac{2}{\pi}$$

From the other side, the empirical correlation coefficient $r$, which is equivalent to the statistic $U$, under the normal distribution is asymptotically normal

$$\sqrt{n}(r - \rho) \xrightarrow{d} Z \sim N(0, (1 - \rho^2)^2) \qquad [4.28]$$

so

$$\mu_2(\rho) = \rho, \quad \mu_2'(0) = 1$$

If $\rho = 0$ then by the formula for the variance of $r_K$ [4.17] and the convergence [4.28]:

$$\sigma_1^2(0) = \lim_{n \to \infty} n\mathbf{Var}_0(r_K) = \frac{4}{9}, \quad \sigma_2^2(0) = 1$$

By [1.5], the asymptotic relative efficiency of Kendall's correlation coefficient $r_K$ with respect to the empirical correlation coefficient $r$ is

$$e(r_K, r) = \frac{(\frac{2}{\pi})^2 : \frac{4}{9}}{1 : 1} = \frac{9}{\pi^2}$$

$\triangle$

### 4.3.4. *Normal scores independence test*

It was proved that if the distribution of $(X, Y)$ is normal then the asymptotic relative efficiency (ARE) of the independence test based on Kendall's rank correlation coefficient (or on the equivalent Spearman's rank correlation coefficient) with respect to the test based on Pearson's empirical correlation coefficient is close to 1. The question arises: can a rank test having $ARE = 1$ be found?

The answer is positive if we use specially chosen ranks functions.

Arrange the pairs $(X_i, Y_i)^T$ so that $Y_1, ..., Y_n$ are arranged in increasing order and replace the $X$ and $Y$ values by ranks.

The lines $(1, \ldots, n)$ and $(R_1, \ldots, R_n)$ are obtained. Denote by

$$E_i = \mathbf{E}(U_{(i)}), \quad i = 1, 2 \ldots, n \qquad [4.29]$$

the means of the order statistics $U_{(i)}$ of the standard normal distribution $N(0, 1)$. The means $E_i$ are called the *normal scores*.

Instead of Spearman's correlation coefficient, i.e. the empirical correlation coefficient between $(1, \ldots, n)$ and $(R_1, \ldots, R_n)$, we can consider the empirical correlation coefficient between $(1, \ldots, n)$ and $(E_{R_1}, \ldots, E_{R_n})$

$$r_{ns} = \frac{\sum_{i=1}^{n}(E_{R_i} - \bar{E})(i - \frac{n+1}{2})}{\sqrt{\sum_{i=1}^{n}(E_{R_i} - \bar{E})^2 \sum_{i=1}^{n}(i - \frac{n+1}{2})^2}} =$$

$$\frac{\frac{1}{n}\sum_{i=1}^{n} iE_{R_i} - \frac{n+1}{2}\bar{E}}{\sqrt{\frac{n^2-1}{12}\frac{1}{n}\sum_{i=1}^{n}(E_{R_i} - \bar{E})^2}} \qquad [4.30]$$

**Normal scores independence test:** independence hypothesis $H_0$ is rejected with a significance level not greater than $\alpha$ if

$$r_{ns} \leq c_1 \quad \text{or} \quad r_{ns} \geq c_2 \qquad [4.31]$$

where $c_1$ is the maximum and $c_2$ is the minimum possible value of $r_S$ satisfying the inequalities

$$\mathbf{P}\{r_{ns} \leq c_1\} \leq \alpha/2, \quad \mathbf{P}\{r_{ns} \geq c_2\} \leq \alpha/2$$

The probability distribution of the random variable $r_{ns}$, hence the critical values $c_1$ and $c_2$, can be found using the probability distribution of the random vector $(R_1, \ldots, R_n)^T$ given in [4.2] and the expressions of $E_i$ as functions of $i$.

It can be shown that if the probability distribution of the random vector $(X, Y)^T$ is normal then the ARE of the independence test based on the statistic $r_{ns}$ with respect to the test based on Pearson's empirical correlation coefficient is equal to 1.

## 4.4. Randomness tests

Suppose that the coordinates of the sample $\mathbf{X} = (X_1, ..., X_n)^T$ are independent absolutely continuous random variables. Denote by $F_i(x)$ the cdf of the random variable $X_i$.

**Randomness hypothesis:**

$$H_0 : F_1(x) \equiv F_2(x) \equiv \cdots \equiv F_n(x)$$

This hypothesis means that $\mathbf{X}$ is a simple sample of a random variable $X$, i.e. all random variables are iid.

If hypothesis $H_0$ holds then the rank vector $(R_1, \ldots, R_n)^T$ of the sample $(X_1, \ldots, X_n)^T$ and the coordinates of this rank vector have the probability distributions given in section 4.1 (see [4.1])

$$\mathbf{P}\{(R_1, ..., R_n) = (j_1, ..., j_n)\} = \frac{1}{n!}, \quad \mathbf{P}\{R_i = j\} = \frac{1}{n}$$

for all permutations $(j_1, ..., j_n)$ of $(1, ..., n)$ and for all $i, j = 1, \ldots n$.

Under the monotonic alternatives

$$\bar{H}_1 : F_1(x) \leq F_2(x) \leq \cdots \leq F_n(x)$$

or

$$\bar{H}_2 : F_1(x) \geq F_2(x) \geq \cdots \geq F_n(x)$$

meaning an increasing or decreasing trend, respectively, the ranks tend to arrange in increasing or decreasing order. So the distribution of the rank $R_i$ depends on the number $i$.

### 4.4.1. *Kendall's and Spearman's randomness tests*

Let $r_S$ and $r_K$ be Spearman's and Kendall's rank correlation coefficients using the vector $(1, 2. \ldots, n)^T$ and the vector of ranks $(R_1, \ldots, R_n)^T$. If the hypothesis $H_0$ holds then the probability distributions of these coefficients are the same as in sections 4.1 and 4.2 and their values are scattered around zero. If the alternatives $\bar{H}_1$ or $\bar{H}_2$ are true then the rank vector $(R_1, \ldots, R_n)^T$ tends to take values near to $(1, \ldots, n)^T$ and $(n, \ldots, 1)^T$, respectively, so both statistics $r_S$ and $r_K$ tend to take values near $+1$ or $-1$, respectively.

**Spearman's (Kendall's) randomness test:** the randomness hypothesis $H_0$ is rejected against the two-sided alternative $\bar{H}_1 \cup \bar{H}_2$ if $|r_S| \geq r_{S,\alpha/2}$ (respectively $|r_K| \geq r_{K,\alpha/2}$), where $r_{S,\alpha/2}$ (respectively $r_{K,\alpha/2}$) is $\alpha/2$ critical value of the statistic $r_S$ (respectively $r_K$).

Let us find the ARE of Kendall's randomness test with respect to the optimal parametric test in the case of a normal distribution when a specified trend alternative is chosen.

Suppose that under the alternative the distribution of the random variables $X_i$ is defined by the following regression model:

$$X_i = \beta_0 + i\beta + e_i, \quad i = 1, 2 \ldots, n \qquad [4.32]$$

where $e_i$ are iid random variables $e_i \sim N(0, 1)$.

The randomness hypothesis is equivalent to the parametric hypothesis $H : \beta = 0$.

From regression analysis it is known that the uniformly most powerful unbiased test is based on the estimator

$$\hat{\beta} = \frac{\sum_i (X_i - \bar{X})(i - \bar{i})}{\sum_i (i - \bar{i})^2} \sim N\left(\beta, \frac{12}{n(n^2 - 1)}\right) \qquad [4.33]$$

of the parameter $\beta$.

**Theorem 4.6.** *The asymptotic relative efficiency of the Kendall's randomness test with respect to the test based on the statistic $\hat{\beta}$ is*

$$e(r_K, \hat{\beta}) = \left(\frac{3}{\pi}\right)^{1/3} \approx 0.985$$

**Proof.** [4.33] implies

$$n^{3/2}(\hat{\beta} - \beta)/\sqrt{12} \xrightarrow{d} Y \sim N(0, 1)$$

So the functions in the expression for ARE corresponding to the parametric test based on $\hat{\beta}$ are

$$\mu_2(\beta) = \beta, \quad \sigma_2(\beta) = \sqrt{12}, \quad \mu_2'(0) = 1, \quad \sigma_2(0) = \sqrt{12}, \quad \delta = 3/2$$

Let us consider the randomness test based on Kendall's correlation coefficient

$$r_K = 1 - 4I_n/(n(n-1))$$

where $I_n$ is the number of inversions. Let us find the mean of $I_n$ under alternative [4.32]. We obtain

$$\mathbf{E}_\beta(I_n) = \mathbf{E}_\beta\left(\sum_{i<j} h_{ij}\right) = \sum_{i<j} \mathbf{E}_\beta(h_{ij})$$

Model [4.32] implies $X_i - X_j \sim N(\beta(i-j), 2)$. So

$$\mathbf{E}_\beta(h_{ij}) = \mathbf{P}_\beta\{X_i - X_j > 0\} =$$

$$= \frac{1}{2\sqrt{\pi}} \int_0^\infty \exp\{-\frac{1}{4}(t - \beta(i-j))^2\}dt = 1 - \Phi\left(\frac{-\beta(i-j)}{\sqrt{2}}\right)$$

Since

$$\frac{\partial}{\partial\beta}(\mathbf{E}_\beta(h_{ij}))|_{\beta=0} = \varphi(0)\frac{i-j}{\sqrt{2}} = \frac{i-j}{2\sqrt{\pi}}$$

and

$$\frac{\partial}{\partial \beta} \mathbf{E}_\beta(I_n)|_{\beta=0} = \frac{1}{2\sqrt{\pi}} \sum_{i<j}(i-j) = -\frac{n(n^2-1)}{12\sqrt{\pi}}$$

we shall use the statistic $r_K^* = r_K/(n+1)$ (which is equivalent to the statistic $r_K$) to find $\mu_1'(0)$. We obtain

$$\mu_1'(0) = \frac{\partial}{\partial \beta} \mathbf{E}_\beta(r_K^*)|_{\beta=0} = \frac{1}{3\sqrt{\pi}} \qquad [4.34]$$

In Theorem 4.4 it was proved that

$$\mathbf{Var}_\beta(I_n)|_{\beta=0} = \frac{n(n-1)(2n+5)}{72}$$

so

$$\sigma_1^2(0) = \lim_{n\to\infty} \mathbf{Var}_\beta(n^{3/2}r_k^*)|_{\beta=0} =$$

$$\lim_{n\to\infty} \frac{n^3}{(n+1)^2} \frac{16}{n^2(n-1)^2} \frac{n(n-1)(2n+5)}{72} = \frac{4}{9}$$

We obtain

$$\mu_1'(0) = \frac{1}{3\sqrt{\pi}}, \quad \sigma_1(0) = \frac{2}{3}$$

Since $\delta = 3/2$, the asymptotic relative efficiency of $r_K$ with respect to the test based on the statistic $\hat{\beta}$ is

$$e(r_K, \hat{\beta}) = \left(\frac{\frac{1}{3\sqrt{\pi}}\frac{3}{2}}{\frac{1}{\sqrt{12}}}\right)^{2/3} = \left(\frac{3}{\pi}\right)^{1/3}$$

$\triangle$

**Example 4.3.** The data on deviations of the temperatures for all days in November 2008 from the mean temperatures of the corresponding days (mean temperatures are obtained from many years of observation) are given in the following table.

| Day | 1 | 2 | 3 | 4 | 5 | 6 | 7 | 8 | 9 | 10 |
|---|---|---|---|---|---|---|---|---|---|---|
| Deviation | 12 | 13 | 12 | 11 | 5 | 2 | −1 | 2 | −1 | 3 |
| Day | 11 | 12 | 13 | 14 | 15 | 16 | 17 | 18 | 19 | 20 |
| Deviation | 2 | −6 | −7 | −7 | −12 | −9 | 6 | 7 | 10 | 6 |
| Day | 21 | 22 | 23 | 24 | 25 | 26 | 27 | 28 | 29 | 30 |
| Deviation | 1 | 1 | 3 | 7 | −2 | −6 | −6 | −5 | −2 | −1 |

Test the hypothesis that the observed deviations are realizations of iid random variables.

Write the values of the ranks in the following table.

| Day | 1 | 2 | 3 | 4 | 5 | 6 | 7 | 8 | 9 | 10 |
|-----|------|------|------|------|-----|-----|------|------|-----|------|
| Ranks | 28.5 | 30 | 28.5 | 27 | 21 | 17 | 12 | 17 | 12 | 19.5 |
| Day | 11 | 12 | 13 | 14 | 15 | 16 | 17 | 18 | 19 | 20 |
| Ranks | 17 | 6 | 3.5 | 3.5 | 1 | 2 | 22.5 | 24.5 | 26 | 22.5 |
| Day | 21 | 22 | 23 | 24 | 25 | 26 | 27 | 28 | 29 | 30 |
| Ranks | 14.5 | 14.5 | 19.5 | 24.5 | 9.5 | 6 | 6 | 8 | 9.5 | 12 |

Using the SPSS package we obtain $r_S = -0.425$ and $r_K = -0.321$. The $P$-values of the Kendall's and Spearman's randomness tests are $pv_K = 0.01424$ and $pv_S = 0.01932$, respectively. If the significance level is greater than 0.01932 then both tests reject the randomness hypothesis.

### 4.4.2. Bartels–Von Neuman randomness test

Another way to see the trend in a sequence of observations is to consider the differences in ranks of adjacent elements. If a trend exists then adjacent elements tend to have small rank differences. The *Bartels–Von Neuman rank test statistic* has the form

$$T_B = \sum_{i=1}^{n-1} (R_{i+1} - R_i)^2. \qquad [4.35]$$

This statistic takes values between $n-1$ and $(n-1)(n^2+n-3)/3$ if $n$ is even, and between $n-1$ and $[(n-1)(n^2+n-3)/3]-1$ if $n$ is odd.

**Bartels–Von Neuman rank test.** The randomness hypothesis is rejected if $T_B \leq c$; here $c$ is the largest number verifying the condition $\mathbf{P}\{T_{BN} \leq c\} \leq \alpha$.

If $4 \leq n \leq 10$ then the $P$-values $pv = \mathbf{P}\{T_B \leq c\}$ for all $T_B$ possible realizations $c$ are given in [GIB 09].

If $n \geq 10$ then the normalized statistic

$$T_{nB} = \frac{T_B}{\sum_{i=1}^{n}(R_i - (n+1)/2)^2}$$

is used. If all ranks are different then the denominator is equal to $n(n^2 - 1)/12$.

If $10 \leq n \leq 100$ then the approximate critical values of the statistic $T_{nB}$ are obtained by approximating its distribution by the beta distribution, and they are given in [GIB 09].

If $n > 100$ then the distribution of the statistic $T_{nB}$ is approximated by the normal distribution $N(2, \sigma_n^2)$; here

$$\sigma_n^2 = \frac{4(n-2)(5n^2 - 2n - 9)}{5n(n+1)(n-1)^2}$$

**Asymptotic Bartels–Von Neuman rank test.** The randomness hypothesis is rejected with an asymptotic significance level $\alpha$ if

$$Z_n = \frac{T_{nB} - 2}{\sigma_n} \leq -z_\alpha$$

.

**Example 4.4.** (continuation of Example 4.3) Test the randomness hypothesis using the data in exercise 4.3 and the Bartels–Von Neuman test.

Ranks and rank differences are given in the following table.

| Day | 1 | 2 | 3 | 4 | 5 | 6 |
|---|---|---|---|---|---|---|
| Ranks | 28.5 | 30 | 28.5 | 27 | 21 | 17 |
| Difference of ranks | | 1.5 | −1.5 | −1.5 | −6 | −4 |
| Day | 7 | 8 | 9 | 10 | 11 | 12 |
| Ranks | 12 | 17 | 12 | 19.5 | 17 | 6 |
| Difference of ranks | −5 | 5 | −5 | 7.5 | −2.5 | −11 |
| Day | 13 | 14 | 15 | 16 | 17 | 18 |
| Ranks | 3.5 | 3.5 | 1 | 2 | 22.5 | 24.5 |
| Difference of ranks | −2.5 | 0 | −2.5 | 1 | 20.5 | 2 |
| Day | 19 | 20 | 21 | 22 | 23 | 24 |
| Ranks | 26 | 22.5 | 14.5 | 14.5 | 19.5 | 24.5 |
| Difference of ranks | 1.5 | −3.5 | −8 | 0 | 5 | 5 |
| Day | 25 | 26 | 27 | 28 | 29 | 30 |
| Ranks | 9.5 | 6 | 6 | 8 | 9.5 | 12 |
| Difference of ranks | −15 | −3.5 | 0 | 2 | 1.5 | 2.5 |

We obtain:

$$T_B = \sum_{i=1}^{29}(R_{i+1} - R_i)^2 = 1133.25, \quad \sum_{i=1}^{30}(R_i - 15.5)^2 = 2238$$

$$T_{nB} = 1133.25/2238 = 0.506367$$

In [GIB 09], we find that the 0.005, 0.01, 0.05, 0.1 critical values of the statistic $T_{nVN}$ are 1.11, 1.19, 1.41 and 1.54, respectively. The value of the statistic $T_{nB}$ is considerably smaller. The hypothesis is rejected. Let us find the asymptotic $P$-value. The value of the statistic $Z$ is −4.1928. The asymptotic $P$-value is $pv_a = \Phi(-4.1928) = 0.0000138$. The hypothesis is rejected.

Note that in the case of this example the $P$-value from the Bartels–Von Neuman test is considerably smaller than the $P$-value from the Kendall's and Spearman's tests (see Example 4.3).

## 4.5. Rank homogeneity tests for two independent samples

Suppose that $\mathbf{X} = (X_1, ..., X_m)^T$ and $\mathbf{Y} = (Y_1, ..., Y_n)^T$ are two independent simple samples of continuous random variables $X \sim F$ and $Y \sim G$. Let us consider the hypothesis stating that the cumulative distribution functions $F$ and $G$ coincide

$$H_0 : F(x) = \mathbf{P}\{X \leq x\} \equiv \mathbf{P}\{Y \leq x\} = G(x) \qquad [4.36]$$

### 4.5.1. *Wilcoxon (Mann–Whitney–Wilcoxon) rank sum test*

Let us consider the following alternative.

**Location alternative:**

$$H_1 : \quad \text{there exists } \theta \neq 0 : \quad G(x) = F(x - \theta) \quad \text{for all} \quad x \in \mathbf{R}$$
$$[4.37]$$

**Wilcoxon rank sum test statistic.** The Wilcoxon rank sum test is based on the sum of ranks of the elements of the first sample

$$W = \sum_{i=1}^{m} R_i$$

in the unified sample $(X_1, ..., X_m, Y_1, ..., Y_n)^T$, where $R_i$ is the rank of $X_i$ in the unified sample.

**Distribution of the test statistic.** If $H_0$ holds then the distribution of the statistic $W$ does not depend on unknown parameters but depends on the sample sizes $m$ and $n$, because, using [4.1], we obtain

$$\mathbf{P}\{(R_1, ..., R_m) = (j_1, ..., j_m)\} = \frac{n!}{(m+n)!} \qquad [4.38]$$

for all $(j_1, ..., j_m)$ obtained from $m$ different elements of the set $\{1, 2...., m + n\}$. The minimum value of the statistic $W$ is

$$w_1 = 1 + \cdots + m = m(m + 1)/2$$

and the maximuml value is

$$w_2 = (n + 1) + (n + 2) + \cdots + (n + m) = m(2n + m + 1)/2$$

So, for all $k = w_1, w_1 + 1, \ldots, w_2$

$$\mathbf{P}\{W = k\} = N_k \frac{n!}{(m + n)!}$$

where $N_k$ is the number of vectors $(j_1, ..., j_m)$ satisfying the condition $j_1 + \cdots + j_m = k$.

**Mann–Whitney test statistic.** Suppose that $U$ is the number of cases where the elements of the first sample are greater than the elements of the second sample

$$U = \sum_{i=1}^{m} \sum_{j=1}^{n} h_{ij}, \quad h_{ij} = \begin{cases} 1, & \text{if } X_i > Y_j, \\ 0, & \text{if } X_i < Y_j. \end{cases} \qquad [4.39]$$

The statistics $W$ and $U$ are closely related. Indeed, let $i_1, \ldots, i_m$ be the indices of the elements of the sample $(X_1, \ldots, X_m)^T$ satisfying the inequalities $R_{i_1} < \cdots < R_{i_m}$. There are $R_{i_l} - 1$ elements to the left from $X_{i_l}$ in the unified ordered sample, $l - 1$ from the first sample and $R_{i_l} - l$ from the second sample, so

$$U = \sum_{i=1}^{m} \sum_{j=1}^{n} h_{ij} = \sum_{l=1}^{m} (R_{i_l} - l) = W - m(m + 1)/2 \qquad [4.40]$$

Under the alternative $\theta > 0$, the elements of the second sample tend to take greater values than the elements of the first sample, and if $\theta < 0$, the statistic $W$ tends to take greater values.

**Wilcoxon rank sum test:** the zero hypothesis is rejected against the two-sided alternative with a significance level not greater than $\alpha$ if

$$W \le c_1 \quad \text{or} \quad W \ge c_2$$

where $c_1$ is the maximum and $c_2$ is the minimum natural number verifying the inequalities

$$\sum_{k=w_1}^{c_1} \mathbf{P}\{W = k|H\} \le \alpha/2. \quad \sum_{i=c_2}^{w_2} \mathbf{P}\{W = k|H\} \le \alpha/2$$

For small $m$ and $n$, the critical values of the statistic $W$ have been tabulated (see [BOL 83]).

If the alternatives are one-sided, $\theta > 0$ or $\theta < 0$, the critical regions are one-sided and of the form $W \ge d$ or $W \le c$, respectively; the critical values $c$ and $d$ are found analogously to be $c_1$ and $c_2$ replacing $\alpha/2$ by $\alpha$.

**Large samples.** If $m$ and $n$ are large then the normal approximation of the distribution of the random variable $W$ is used. Set $N = m + n$. From [4.3], the means and variances of the sum of ranks $W$ are

$$\mathbf{E}(W) = \frac{m(N+1)}{2}, \mathbf{Var}(W) = \sum_{j=1}^{m} \mathbf{Var}(R_j) + \sum\sum_{i \ne j} \mathbf{Cov}(R_i, R_j) =$$

$$m\frac{N^2 - 1}{12} - m(m-1)\frac{N+1}{12} = \frac{mn(N+1)}{12} \qquad [4.41]$$

Set

$$Z_{m,n} = \frac{W - \mathbf{E}(W)}{\sqrt{\mathbf{Var}(W)}} = \frac{U - \mathbf{E}(U)}{\sqrt{\mathbf{Var}(U)}}$$

**Theorem 4.7.** *If the probability distribution of the random variables $X$ and $Y$ is absolutely continuous, $N \to \infty$, $m/N \to$*

$p \in (0, 1)$, *then under the zero hypothesis*

$$Z_{m,n} \overset{d}{\to} Z \sim N(0, 1)$$

**Proof.** Denote by $S_N$ the number of inversions in the unified sample of size $N = m + n$. This is obtained by comparing all $N(N+1)/2$ observed pairs. Subtracting from $S_N$ the numbers of inversions $S'_m$ and $S''_n$ which are obtained by comparing elements of the first and the second the sample, respectively, gives the number of inversions $U$ obtained by comparing the elements of the first and the second sample

$$S_N = S'_m + S''_n + U, \quad W = U + \frac{m(m+1)}{2} = S_N - S'_m - S''_n + \frac{m(m+1)}{2}$$
$$[4.42]$$

Under the hypothesis, the random variables $S'_m$, $S''_n$ and $U$ are independent. By Theorem 4.4 the random variables $S_N$, $S'_m$, $S''_n$ are asymptotically normal. We obtain

$$\frac{S_N - \mathbf{E}S_N}{\sqrt{\mathbf{Var}(S_N)}} = \frac{S'_m + S''_n - \mathbf{E}(S'_m + S''_n)}{\sqrt{\mathbf{Var}(S'_m + S''_n)}} \sqrt{\frac{\mathbf{Var}(S'_m + S''_n)}{\mathbf{Var}(S_N)}} +$$

$$\frac{U - \mathbf{E}U}{\sqrt{\mathbf{Var}(U)}} \sqrt{\frac{\mathbf{Var}(U)}{\mathbf{Var}(S_N)}} \qquad [4.43]$$

Since

$$\frac{\mathbf{Var}(S'_m + S''_n)}{\mathbf{Var}(S_N)} \to 1 - 3pq, \qquad \frac{\mathbf{Var}(U)}{\mathbf{Var}(S_N)} \to 3pq, \quad q = 1 - p$$

the first term of the right-hand side of equalities [4.42] weakly converges to the random variable $V_1 \sim N(0, 1 - 3pq)$, and the first term of the left-hand side weakly converges to the random variable $V \sim N(0, 1)$, so the second term of the right-hand side converges to $V_2 \sim N(0, 3pq)$, because the terms of the right-hand side are independent. So

$$Z_{m,n} = \frac{U - \mathbf{E}U}{\sqrt{\mathbf{Var}(U)}} \xrightarrow{d} Z \sim N(0, 1)$$

$\triangle$

**Asymptotic Wilcoxon's rank sum test:** if $m$ and $n$ are large then the hypothesis $H_0$ is rejected against the two-sided alternative with an approximate significance level $\alpha$ if

$$|Z_{m,n}| > z_{\alpha/2}$$

If coinciding values exist then (see Theorem 4.2)

$$\mathbf{V}(W) = \sum_{j=1}^{m} \mathbf{Var}(R_j) + \sum\sum_{i \neq j} \mathbf{cov}(R_i, R_j) = m(\frac{N^2 - 1}{12} - \frac{\mathbf{E}T}{12N}) +$$

$$m(m-1)(-\frac{N+1}{12} + \frac{\mathbf{E}T}{12N(N-1)}) = \frac{mn(N+1)}{12}(1 - \frac{\mathbf{E}T}{N^3 - N})$$

where

$$T = \sum_{i=1}^{k} T_i, \quad T_i = (t_i^3 - t_i)$$

$k$ is the number of groups with coinciding values in the unified sample and $t_i$ is the size of the $i$-th group.

So, if $m$ and $n$ are not small then the statistic $Z_{m,n}$ is modified: the statistic

$$Z_{m,n}^* = \frac{Z_{m,n}}{\sqrt{1 - T/(N^3 - N)}}$$

is used.

**Modified asymptotic Wilcoxon rank sum test:** if $m$ and $n$ are large then under the two-sided alternative the hypothesis $H_0$ is rejected with an approximate significance level $\alpha$ if

$$|Z_{m,n}^*| > z_{\alpha/2}$$

**Example 4.5.** (continuation of Example 3.2). Using the data in Example 3.2, use the Wilcoxon–Mann–Whitney rank sum test to verify the hypothesis that fungicides have no influence on the percentage of infections.

The sample sizes are $m = n = 7$, the size of the unified sample is $N = m+n = 14$. Ordered unified sample (the number of the sample is given in parentheses):

| 1 | 2 | 3 | 4 | 5 | 6 | 7 |
|---|---|---|---|---|---|---|
| 0.75(1) | 1.76(1) | 2.46(1) | 4.88(1) | 5.10(1) | 5.68(2) | 5.68(2) |
| 8 | 9 | 10 | 11 | 12 | 13 | 14 |
| 6.01(1) | 7.13(1) | 11.63(2) | 16.30(2) | 21.46(2) | 33.30(2) | 44.20(2) |

The ranks:

| 1(1) | 2(1) | 3(1) | 4(1) | 5(1) | 6.5(2) | 6.5(2) |
|------|------|------|------|------|--------|--------|
| 8(1) | 9(1) | 10(2) | 11(2) | 12(2) | 13(2) | 14(2) |

The sum of the ranks of the first sample (Wilcoxon statistics) is

$$W = 1 + 2 + 3 + 4 + 5 + 8 + 9 = 32$$

The sum of the ranks $N(N + 1)/2 - W = 105 - 32 = 73$ of the second sample is considerably greater, so it may be expected that using fungicides increases the percentage of infections.

Mann–Whitney statistic

$$U = W - \frac{m(m + 1)}{2} = 32 - \frac{7 \cdot 8}{2} = 4$$

Since

$$\mathbf{E}W = m(N + 1)/2 = (7 \cdot 15)/2 = 52.5$$

$$\mathbf{Var}(W) = mn(N + 1)/12 = (7 \cdot 7 \cdot 15)/12 = 61.25$$

we have

$$Z_{m,n} = \frac{W - m(N+1)/2}{\sqrt{mn(N+1)/12}} = \frac{32 - 52.5}{\sqrt{61,25}} \approx -2.619394$$

There is one pair of coinciding ranks, $k = 1$ and $t_1 = 2$, so $T_1 = (2^3 - 2) = 6$, and

$$1 - \frac{\sum_{i=1}^{k} T_i}{n^3 - n} = 1 - \frac{6}{14^3 - 14} = 0.99782$$

The value of the modified statistic is

$$Z_{m,n}^* = Z_{m,n}/\sqrt{0.99782} \approx -2.6223$$

Using the SPSS package we obtain the exact $P$-value $pv = 0.006410$. The asymptotic $P$-value is

$$pv_a = 2(1 - \Phi(2.6223)) \approx 0.008734$$

The zero hypothesis is rejected for any significance level greater than $0.008734$.

For the given data, detection of the difference between two distributions was better using the Wilcoxon's rank sum test than using the two-sample Kolmogorov–Smirnov test, but worse than using the two-sample Cramér–von-Mises test.

Since the sum of ranks of the first sample is smaller and the value of the statistic $Z_{m,n}^*$ is negative, we can expect that using pesticides increases the risk of infection. If it is known that using pesticides cannot decrease the risk of infection then the alternative is one-sided

$$H_1: \quad \exists \theta < 0: \quad G(x) = F(x - \theta) \quad \text{for all} \quad x \in \mathbf{R}$$

Under this alternative, the distributions of $Y$ and $X - \theta$ coincide, so $Y$ (the percentage of infections when pesticides are used) tends to take greater values than $X$ (the percentage of infections when pesticides are not used).

In this case the hypothesis is rejected if $W \leq c$. The hypothesis is rejected by the asymptotic test if $Z_{m,n}^* < -z_\alpha$. Using the SPSS package we obtained the $P$-value $pv = 0.0032051$. The asymptotic $P$-value is

$$pv_a = \Phi(-2.6223)) \approx 0.004367$$

The hypothesis is rejected with a smaller $P$-value than in the case of the two-sided alternative.

### 4.5.2. Power of the Wilcoxon rank sum test against location alternatives

Let us find an approximation of the power of the Wilcoxon rank sum test when samples are large. As above, let us consider the location alternative $\bar{H}$ (see [4.37]). The homogeneity hypothesis is equivalent to the hypothesis $H_0$ : $\theta = 0$, signifying equality of the location parameters.

Denote by $f(x)$ and $f(x - \theta)$ the pdfs of the variables $X$ and $Y$, respectively.

Under the alternative, the distribution of the random variables $h_{ij}$ defined by [4.39] is found using the formula of complete probabilities

$$p_1(\theta) = \mathbf{P}\{h_{11} = 1\} = \mathbf{P}\{X_1 > Y_1\} = \int_{-\infty}^{\infty} F(x - \theta)f(x)dx$$

$$p_2(\theta) = \mathbf{P}\{h_{11} = 1,\ h_{12} = 1\} = \mathbf{P}\{X_1 > Y_1, X_1 > Y_2\} =$$

$$\int_{-\infty}^{\infty} F^2(x - \theta)f(x)dx$$

$$p_3(\theta) = \mathbf{P}\{h_{11} = 1, h_{21} = 1\} = \mathbf{P}\{X_1 > Y_1, X_2 > Y_1\} =$$

$$\int_{-\infty}^{\infty} [1 - F(x)]^2 f(x - \theta)dx \qquad [4.44]$$

Using expression [4.40] we obtain

$$\mu(\theta) = \mathbf{E}(U) = mn\, p_1(\theta)$$

$$\sigma^2(\theta) = \mathbf{Var}(U) = mn\mathbf{Var}(h_{11}) + mn(n-1)\mathbf{Cov}(h_{11}, h_{12}) +$$

$$nm(m-1)\mathbf{Cov}(h_{11}, h_{21}) =$$

$$mn\,[p_1(\theta) - p_1^2(\theta) + (n-1)(p_2(\theta) - p_1^2(\theta)) + (m-1)(p_3(\theta) - p_1^2(\theta))]$$

If $m$ and $n$ are large then by the CLT theorem the distribution of the random variable $(U - \mu(\theta))/\sigma(\theta)$ is approximated by the standard normal law. So the power of the test is approximated as follows

$$\beta(\theta) = \mathbf{P}_\theta\{|\frac{U - \mu(0)}{\sigma(0)}| > z_{\alpha/2}\} =$$

$$\mathbf{P}_\theta\{\frac{U - \mu(\theta)}{\sigma(\theta)} > \frac{\mu(0) - \mu(\theta) + \sigma(0)z_{\alpha/2}}{\sigma(\theta)}\} +$$

$$\mathbf{P}_\theta\{\frac{U - \mu(\theta)}{\sigma(\theta)} < \frac{\mu(0) - \mu(\theta) - \sigma(0)z_{\alpha/2}}{\sigma(\theta)}\} \approx$$

$$1 - \Phi\left(\frac{\mu(0) - \mu(\theta) + \sigma(0)z_{\alpha/2}}{\sigma(\theta)}\right) + \Phi\left(\frac{\mu(0) - \mu(\theta) - \sigma(0)z_{\alpha/2}}{\sigma(\theta)}\right)$$

Note that the functions $p_1, p_2$ and $p_3$ and the power depend not only on the parameter $\theta$ but also on the cdf $F$.

By specifying the class of the functions $F$, approximate formulas for the power, depending only on scalar parameters, can be obtained.

Suppose that the function $F(x)$ belongs to the location-scale family

$$\{F_0((x - \mu)/\sigma), \mu \in \mathbf{R}, \sigma > 0\}$$

where $F_0(y)$ is a known function. Denote by $f_0$ the cdf of the hypothesis $H : \eta = 0$, where $\eta = \theta/\sigma$. In this case, the functions $p_1, p_2$ and $p_3$ depend only on the parameter $\eta$

$$p_1(\eta) = \int_{-\infty}^{\infty} F_0(y - \eta)f_0(y)dy, \quad p_2(\eta) = \int_{-\infty}^{\infty} F_0^2(y - \eta)f_0(y)dy$$

$$p_3(\eta) = \int_{-\infty}^{\infty} [1 - F_0(y)]^2 f_0(y - \eta)dy$$

So

$$\beta(\eta) \approx 1 - \Phi \left( \frac{\mu(0) - \mu(\eta) + \sigma(0)z_{\alpha/2}}{\sigma(\eta)} \right) +$$

$$\Phi \left( \frac{\mu(0) - \mu(\eta) - \sigma(0)z_{\alpha/2}}{\sigma(\eta)} \right)$$

### 4.5.3. ARE of the Wilcoxon rank sum test with respect to the asymptotic Student's test

Under the location alternative [4.38], the equality $\mathbf{E}Y_j = \mathbf{E}X_i + \theta$ holds and the homogeneity hypothesis, being equivalent to the hypothesis $H_0 : \theta = 0$, is equivalent to the means equality hypothesis $H_0 : \mu_1 = \mu_2$, where $\mu_1 = \mathbf{E}X_i$, $\mu_2 = \mathbf{E}Y_i$. The last hypothesis can be verified using the asymptotic Student's test.

**Asymptotic Student's test statistic.** Under $H_0$, $X_i \sim F$, $Y_j \sim F$, so

$$\mathbf{E}X_i = \mathbf{E}Y_j, \quad \mathbf{Var}(X_i) = \mathbf{Var}(Y_j) =: \tau^2, \quad \mathbf{E}(\bar{X} - \bar{Y}) = 0$$

$$\mathbf{Var}(\bar{X} - \bar{Y}) = \tau^2 (\frac{1}{m} + \frac{1}{n})$$

Set

$$S^2 = ((m - 1)s_1^2 + (n - 1)s_2^2)/(m + n - 2)$$

where $s_1^2$ and $s_2^2$ are the sample variances of the first and the second sample, respectively.

If $N = m + n \to \infty$, $m/N \to p \in (0, 1)$ then

$$t = \frac{\bar{X} - \bar{Y}}{S\sqrt{\frac{1}{m} + \frac{1}{n}}} \xrightarrow{d} Z \sim N(0, 1) \qquad [4.45]$$

**Asymptotic Student's test:** if $m$ and $n$ are large then the hypothesis $H_0$ is rejected with an approximate significance level $\alpha$ when $|t| > z_{\alpha/2}$.

In the case of the normal distribution, i.e. when $F_0 = \Phi$, the statistic $t$ is used when samples are small. In such cases, it has a Student's distribution with $m + n - 2$ degrees of freedom and the hypothesis $H_0$ is rejected if $|t| > t_{\alpha/2}(m + n - 2)$.

Let us find the ARE of Wilcoxon's rank sum test with respect to Student's test.

**Theorem 4.8.** *If $N \to \infty$, $m/N \to p \in (0,1)$ then the ARE of the Wilcoxon rank sum test with respect to the Student's test is*

$$e(W, t) = 12\tau^2 \left[ \int_{-\infty}^{\infty} f^2(x)dx \right]^2, \quad \tau^2 = \text{Var}(X_i) \qquad [4.46]$$

**Proof.** The Student's test statistic $t$ is asymptotically equivalent (as $N \to \infty$, $m/N \to p \in (0, 1)$) to the normalized statistic $\bar{X} - \bar{Y}$, because $S \xrightarrow{P} \tau$. Under the location alternative $H_1$

$$\frac{\sqrt{N}(\bar{X} - \bar{Y} + \theta)}{\tau\sqrt{1/p + 1/q}} \xrightarrow{d} Z \sim N(0\,1), \quad q = 1 - p$$

So the functions corresponding to the Student's test in the formula of the ARE [1.5] has the form

$$\mu_2(\theta) = -\theta, \quad \sigma_2(\theta) = \tau\sqrt{1/p + 1/q} = \frac{\tau}{\sqrt{pq}}, \quad \mu'_2(0) = -1$$

$$\sigma_2(0) = \frac{\tau}{\sqrt{pq}}$$

By investigating the power of Wilcoxon's rank sum test it was shown that

$$(U - \mu(\theta))/\sigma(\theta) \xrightarrow{d} Z \sim N(0, 1)$$

The statistic $U^* = U/(mn)$ is equivalent to the statistic $U$, so we use the former

$$\frac{\sqrt{N}(U^* - p_1(\theta))}{\sqrt{(p_2(\theta) - p_1^2(\theta))/q + (p_3(\theta) - p_1^2(\theta))/p}} \xrightarrow{d} Z \sim N(0, 1)$$

So, the functions corresponding to Wilcoxon's rank sum test in the ARE formula [1.5] have the form

$$\mu_1(\theta) = p_1(\theta), \quad \sigma_1^2(\theta) = (p_2(\theta) - p_1^2(\theta))/q + (p_3(\theta) - p_1^2(\theta))/p$$

Differentiating with respect to $\theta$, we obtain

$$\mu_1'(\theta) = -\int_{-\infty}^{\infty} f(x - \theta)f(x)dx, \quad \mu_1'(0) = -\int_{-\infty}^{\infty} f^2(x)dx$$

Since $p_1(0) = 1/3$, $p_2(0) = p_3(0) = 1/3$, we have $\sigma_1^2(0) = \frac{1}{12pq}$.

So, from [1.5], the ARE of the Wilcoxon's statistic with respect to the Student's statistic is

$$e(W, t) = \left(\frac{-\int_{-\infty}^{\infty} f^2(x)dx \, 2\sqrt{3pq}}{(-1)\frac{\sqrt{pq}}{\tau}}\right)^2 = 12\tau^2 \left[\int_{-\infty}^{\infty} f^2(x)dx\right]^2$$

$\triangle$

**Location-scale families.** If the function $F(x)$ belongs to the family

$$\{F_0((x - \mu)/\sigma), \mu \in \mathbf{R}, \sigma > 0\}$$

where $F_0(y)$ is a known function, then the ARE does not depend on unknown parameters

$$e(W, t) = 12\tau_0^2 \left[\int_{-\infty}^{\infty} f_0^2(y)dy\right]^2$$

where

$$\tau_0^2 = \mathbf{Var}((X_i - \mu)/\sigma) = \int_{-\infty}^{\infty} y^2 dF_0(y) - (\int_{-\infty}^{\infty} y dF_0(y))^2$$

because

$$\int_{-\infty}^{\infty} f^2(x)dx = \sigma^{-1}\int_{-\infty}^{\infty} f_0^2(x)dx, \quad \tau^2 = \sigma^2\tau_0^2$$

Let us consider several examples.

1) Normal distribution: $F_0 = \Phi$, $f_0 = \varphi$, $\tau_0^2 = 1$

$$\int_{-\infty}^{\infty} \varphi^2(y)dy = \frac{1}{2\pi}\int_{-\infty}^{\infty} \exp\{-x^2\}dx = \frac{1}{2\sqrt{\pi}}$$

hence

$$e(W,t) = \frac{3}{\pi} \approx 0.95$$

2) Uniform distribution:

$$F_0(x) = \begin{cases} 0, & \text{if } x \leq -1, \\ (x+1)/2, & \text{if } x \in (-1,1), \\ 1, & \text{if } x \geq 1, \end{cases}$$

$$f_0(x) = \frac{1}{2}\mathbf{1}_{(-1,1)}(x), \quad \tau_0^2 = 1/3, \quad \int_{-\infty}^{\infty} f_0^2(x)dx = \frac{1}{2}$$

so

$$e(W,t) = 1$$

3) Logistic distribution:

$$F_0(x) = \frac{1}{1+e^{-x}}, \quad f_0(x) = \frac{e^{-x}}{(1+e^{-x})^2}, \quad \tau_0^2 = \frac{\pi^2}{3}$$

$$\int_{-\infty}^{\infty} f_0^2(x)dx = \int_{-\infty}^{\infty} \frac{e^{-2x}}{(1+e^{-x})^4}dx = \int_{0}^{\infty} \frac{y}{(1+y)^4}dy = \frac{1}{6}$$

so

$$e(W,t) = \frac{\pi^2}{9} \approx 1.097$$

4) Extreme values distribution:

$$F_0(x) = 1 - e^{-e^x}, \quad f_0(x) = e^x e^{-e^x}, \quad \tau_0^2 = \frac{\pi^2}{6}$$

$$\int_{-\infty}^{\infty} f_0^2(x)dx = \frac{1}{4}\int_0^{\infty} ye^{-y}dy = \frac{1}{4}$$

so

$$e(W,t) = \frac{\pi^2}{8} \approx 1.23$$

5) Doubly exponential distribution:

$$F_0(x) = \begin{cases} \frac{1}{2}e^x, & \text{if } x \le 0 \\ 1 - \frac{1}{2}e^{-x}, & \text{if } x > 0 \end{cases}$$

$$f_0(x) = \frac{1}{2}e^{-|x|}, \quad \tau_0^2 = 2$$

$$\int_{-\infty}^{\infty} f_0^2(x)dx = \frac{1}{2}\int_0^{\infty} e^{-2x}dx = \frac{1}{4}$$

so

$$e(W,t) = 2$$

**Comment 4.8.** The considered examples show that for some families of distributions the Wilcoxon rank sum test is more efficient than the Student's test. Moreover, the integral in formula [4.46] can take infinite values (see exercise 4.8). So, the value $e(W,t) = \infty$ is possible.

Let us find the distribution giving the minimum asymptotic relative efficiency $e(W,t)$ and find this minimum.

**Theorem 4.9.** *The minimum of the ARE of Wilcoxon's statistic with respect to the Student's statistic is* $\inf_f e(W,t) = 0.864.$

**Proof.** From [4.46], it suffices to minimize

$$\mathbf{E}(f(X)) = \int_{-\infty}^{\infty} f^2(x)dx$$

where $X$ is an absolutely continuous random variable with pdf $f(x)$ and variance $\tau^2 = 1$. Since the statistics $W$ and $t$ do not change when the random variables $X_i$ and $Y_j$ are replaced by the random variables $X_i + \mu$ and $Y_j + \mu$, respectively, where $\mu = \mathbf{E}X_i$, we can suppose that $\mu = 0$. So, from [4.46], we must minimize the integral

$$\mathbf{E}(f(X)) = \int_{-\infty}^{\infty} f^2(x)dx$$

with the conditions

$$\int_{-\infty}^{\infty} f(x)dx = 1, \quad \int_{-\infty}^{\infty} x^2 f(x)dx = 1 \qquad [4.47]$$

Using the Lagrange multiplier method, we must minimize the integral

$$\int_{-\infty}^{\infty} [f(x) - \lambda_1 - \lambda_2 x^2] f(x)dx$$

Since $f(x)$ is non-negative, this integral takes a minimum value when

$$f(x) = \begin{cases} \lambda_1 + \lambda_2 x^2, & \text{if } \lambda_1 + \lambda_2 x^2 \geq 0 \\ 0, & \text{if } \lambda_1 + \lambda_2 x^2 < 0 \end{cases}$$

Putting this expression for the function in [4.47], we obtain a system of two equations for find the Lagrange multipliers $\lambda_1$ and $\lambda_2$. We obtain

$$\lambda_1 = \frac{3}{4\sqrt{5}}, \quad \lambda_2 = -\frac{3}{20\sqrt{5}}$$

$$f(x) = \begin{cases} \frac{3}{4\sqrt{5}} - \frac{3}{20\sqrt{5}} x^2, & \text{if } |x| \leq \sqrt{5}, \\ 0, & \text{if } |x| > \sqrt{5}, \end{cases}$$

$$\left[ \int_{-\sqrt{5}}^{\sqrt{5}} \left( \frac{3}{4\sqrt{5}} - \frac{3}{20\sqrt{5}} x^2 \right)^2 dx \right]^2 = \frac{9}{125}$$

From [4.46]

$$\inf_{f} e(W, t) = 12 \frac{9}{125} = 0.864$$

△

**Comment 4.9.** The performed analysis implies that in the case of a normal distribution, location alternatives and large samples, using Student's test instead of the Wilcoxon's rank sum test, the same power is attained with $5\%$ fewer observations. In the case of the most unfavorable distribution, we can find the test which instead requires $14\%$ fewer observations then Wilcoxon's rank sum test to attain the same power.

On the other hand, there are many distributions where Wilcoxon's rank sum test is more powerful than the Student's test (see examples of logistic and extreme values distributions and exercise 4.8). So, if there are doubts that the distribution of observations is normal, both tests should be tried.

### 4.5.4. *Van der Warden's test*

We saw that in the case of the normal distribution the ARE of the Wilcoxon's rank sum test with respect to the Student's test is approximately $0.95$. Can we find a test based on ranks having the ARE equal to 1? The answer is positive (see [VAN 52]) if instead of the sum of ranks $W = \sum_{i=1}^{m} R_i$ the sum of specified rank functions is used

$$V = v(R_1) + \cdots + v(R_m), \ v(r) = \Phi^{-1}\left(\frac{r}{N+1}\right) \qquad [4.48]$$

The distribution of the random variable $V$ is symmetric with respect to $0$. Indeed, since $\Phi^{-1}(z) = -\Phi^{-1}(1-z)$, we have

$$-V = -\sum_{r=1}^{m} \Phi^{-1}\left(\frac{r}{N+1}\right) = \sum_{r=1}^{m} \Phi^{-1}\left(\frac{N+1-r}{N+1}\right)$$

Under the zero hypothesis, the random vector $(R_1, \ldots, R_m)$ takes each value with the same probability as the random vector $(N + 1 - R_1, \ldots, N + 1 - R_m)$, so the distributions of $V$ and $-V$ coincide.

**Van der Warden test:** under a two-sided alternative the homogeneity hypothesis is rejected with a significance level not greater than $\alpha$ if $|V| > c$, where $c$ is the smallest real number verifying the inequality $\mathbf{P}\{|V| > c|H_0\} \leq \alpha/2$.

If $m$ and $n$ are small, the critical values of the statistic $V$ have been tabulated [BOL 83] and can also be found using statistical packages (SAS, SPSS).

If $N \to \infty, m/N \to p \in (0, 1)$ then the distribution of the random variable $V$ is approximated by the normal $N(0, \sigma_V^2)$, where

$$\sigma_V^2 = \frac{mnQ}{N - 1}, \quad Q = \frac{1}{N} \sum_{r=1}^{N} v^2(r) \qquad [4.49]$$

Set $Z_{m,n} = V/\sigma_V$.

**Asymptotic Van der Warden test:** if $m$ and $n$ are large then under the two-sided alternative the homogeneity hypothesis is rejected with an approximate significance level $\alpha$ if $|Z_{m,n}| > z_{\alpha/2}$.

**Example 4.6.** (continuation of Examples 3.2 and 4.5) Using the data in Example 3.2, use the Van der Warden test to verify the hypothesis that fungicides have no influence on the percentage of infections.

Using the SAS statistical software we have $V = -4.1701$, $pv = 0.0052$, $pv_a = 0.0097$. The hypothesis is rejected.

### 4.5.5. *Rank homogeneity tests for two independent samples under a scale alternative*

Suppose that a scale alternatives against the homogeneity hypothesis is considered

$$H_1 : G(x) = F(\frac{x}{\sigma}), \quad \sigma > 0 \qquad [4.50]$$

If $\sigma > 1$ $(0 < \sigma < 1)$ then the spread of $Y$ values is greater (smaller) than the spread of $X$ values. The medians of the random variables $X$ and $Y$ coincide. Denote the common median by $M$.

**Comment 4.10.** If the region of possible values of the random variables $X$ and $Y$ is $(0, \infty)$ then after the transformations

$$X_i^* = \ln X_i, \quad Y_j^* = \ln Y_j$$

the cdf random variables $X_i^*$ and $Y_j^*$ have the forms

$$F^*(x) = F(e^x) \quad \text{and} \quad G^*(x) = F(\frac{e^x}{\sigma}) = F(e^{x - \ln \sigma}) = F^*(x - \theta)$$

where $\theta = \ln \sigma$. For the transformed variables the alternative is location, so using samples with the elements $\ln X_i$ and $\ln Y_j$ the rank tests discussed above can be applied against the location alternative.

**Comment 4.11.** If the region of possible values of the random variables $X$ and $Y$ is $\mathbf{R}$ then tests based on Wilcoxon's, or any of the other rank tests discused above, used against the location alternatives are not efficient under the scale alternative.

Indeed, suppose that $F(x) = F_0((x - \mu)/\tau)$. Then under the alternative $G(x) = F_0((x - \mu)/(\sigma\tau))$ the distributions of the random variables $X - \mu$ and $(Y - \mu)/\sigma$ coincide, so observations of both samples concentrate around $\mu$. If $\sigma > 1 (0 < \sigma < 1)$ then $X$ observations tend to concentrate closer to (further from) $\mu$ than do $Y$ observations, hence the $X$ observations tend to concentrate in the middle (at the extremities), and $Y$-observations tend to concentrate at the extremities (in the middle) of the unified ordered sample. So, the values of the statistic $W$ (and of the rank statistics considered above) under the alternative are neither small nor large. So the test based on these statistics may not discriminate the zero hypotheses from the alternative.

In the case of scale alternatives, rank functions are chosen as follows.

**Idea of test construction.** Attribute value $s(r)$ to the $r$-th element of the unified ordered sample, where $s$ is a function defined on the set $(1, 2. ..., N)$. Define the statistic

$$S = s(R_1) + \cdots + s(R_m) \qquad [4.51]$$

where, as in the case of Wilcoxon's rank sum test, $R_i$ are ranks of the first sample in the unified ordered sample.

It is natural to attribute the smallest values of $s$ to the smallest and to the largest elements, and the greatest values to the central elements, of the unified ordered sample. Under the alternative $\sigma > 1$ $(0 < \sigma < 1)$, the elements of the first sample concentrate in the middle (at the extremities) of the unified ordered sample, so the sum $S$ takes large (small) values. Under the zero hypothesis, the sum $S$ takes neither large nor small values. So the homogeneity hypothesis is rejected against the one-sided alternative $\sigma > 1$ $(0 < \sigma < 1)$ if $S > c_2$ $(S < c_1)$; here $c_1$ and $c_2$ are the critical values of the statistic $S$ under the zero hypothesis. Similarly, if the alternative is two-sided then the zero hypothesis is rejected if $S < c_1^*$ or $S > c_2^*$. The *Siegel–Tukey* and *Ansari–Bradley* tests are constructed in such a way.

Alternatively, the largest values of $s$ may be attributed to the smallest and to the largest elements, and the smallest values to the central elements of the unified ordered sample. Then the signs of all inequalities are replaced by the opposite signs in the definition of the critical regions. The *Mood* and *Klotz* tests are constructed in such a way.

**Definition of the functions** $s$:

1. Siegel–Tukey test

$$s(1) = 1, \; s(N) = 2. \; s(N-1) = 3, \; s(2) = 4$$

$$s(3) = 5, \; s(N-2) = 6, \; s(N-3) = 7, \; s(4) = 8, \cdots \qquad [4.52]$$

2. Ansari–Bradley test

$$s(1) = 1, \; s(N) = 1, \; s(2) = 2, \; s(N-1) = 2, \cdots \qquad [4.53]$$

3. Mood test

$$s(r) = \left( r - \frac{N+1}{2} \right)^2 \qquad [4.54]$$

4. Klotz test

$$s(r) = \left[ \Phi^{-1} \left( \frac{i}{N+1} \right) \right]^2 \qquad [4.55]$$

Under the homogeneity hypothesis, the first two moments of the Siegel–Tukey, Ansari–Bradley, Mood, and Klotz statistics are

$$\mathbf{E}S_{ZT} = m(N+1)/2, \quad \mathbf{Var}(S_{ZT}) = mn(N+1)/12$$

$$\mathbf{E}S_{AB} = m(N+1)/4, \quad \mathbf{Var}(S_{AB}) = mn(N+1)^2/(48N)$$

$$\mathbf{E}S_M = m(N^2 - 1)/12. \quad \mathbf{Var}(S_M) = mn(N+1)(N^2-4)/180$$

$$\mathbf{E}S_K = \frac{m}{N} \sum_{i=1}^{N} \left[ \Phi^{-1} \left( \frac{i}{N+1} \right) \right]^2$$

$$\mathbf{Var}(S_K) = \frac{mn}{N(N-1)} \sum_{i=1}^{N} \left[ \Phi^{-1} \left( \frac{i}{N+1} \right) \right]^4 - \frac{n}{m(N-1)} [\mathbf{E}S_K]^2$$

Under the zero hypothesis, the distribution of the Siegel–Tukey statistic coincides with the distribution of the Wilcoxon's statistic, and these statistics are asymptotically equivalent to the Ansari–Bradley statistic.

**Large samples.** All statistics are asymptotically normal under the hypothesis $H$

$$Z_{m,n} = \frac{S - \mathbf{E}S}{\sqrt{\mathbf{Var}(S)}} \xrightarrow{d} Z \sim N(0, 1) \text{ as } n \to \infty, \; m/n \to p \in (0, 1)$$

$$[4.56]$$

If $m$ and $n$ are large then the tests are based on the statistics $Z_{m,n}$ and, in dependence on alternatives, the zero hypothesis is rejected if this statistic is larger or smaller than the critical values of the standard normal distribution.

If the distribution is normal and the two populations differ only in variances then the ARE with respect to Fisher's test for the equality of variances is: the Ansari–Bradley and the Siegel–Tukey tests $- 6/\pi^2 \approx 0.608$; the Mood test $- 15/(2\pi^2) \approx 0.760$; the Klotz test $- 1$.

**Example 4.7.** The value of a TV selector's electrical characteristic was measured using two measuring devices. The obtained values of the measurement errors are the following: $m = 10$ measurements, $X_1, ..., X_{10}$, using the first device, and $n = 20$ measurements, $Y_1, ..., Y_{20}$, using the second device. The results are as follows (multiplied by 100).

a) Device of the first type: 2.2722; -1.1502; 0.9371; 3.5368; 2.4928; 1.5670; 0.9585; -0.6089; -1.3895; -0.5112.

b) Device of the second type: 0.6387; -1.8486; -0.1160; 0.6832; 0.0480; 1.2476; 0.3421; -1.5370; 0.6595; -0.7377; -0.0726; 0.6913; 0.4325; -0.2853; 1.8385; -0.6965; 0.0037; -0.3561; -1.9286; 0.4121.

Supposing that both samples are obtained observing independent absolutely continuous random variables $X \sim F(x)$ and $Y \sim G(x)$ with the means $\mathbf{E}X = \mathbf{E}Y = 0$, let us test the hypothesis $H_0 : F(x) \equiv G(x)$ with the one-sided alternative $\bar{H} : G(x) \equiv F(x/\theta), \theta < 1$, which means that the second device is not as exact as the first one.

The values of the test statistics are

$$S_{ZT} = 100, \quad S_{AB} = 52, \quad S_M = 1128.5, \quad S_K = 12.363$$

Using the SAS statistical package we obtain the following $P$-values: 0.0073; 0.0067; 0.0168; 0.0528. Using approximation [4.56] the asymptotic $P$-values are: 0.0082; 0.0068; 0.0154; 0.0462. The homogeneity hypothesis is rejected.

If the distribution of the observed random variables is normal, $X \sim N(0, \sigma_1^2)$ ir $Y \sim N(0, \sigma_2^2)$, then the hypothesis $H_0$ is the parametric hypothesis on the equality of the variances $H_0 : \sigma_1^2 = \sigma_2^2$ with the alternative $\bar{H} : \sigma_1^2 > \sigma_2^2$. Such a hypothesis is tested using the statistic $F = s_1^2/(s_2^2)$, where $s_1^2$ and $s_2^2$ are the estimators of the variances:

$$\hat{\sigma}_1^2 = s_1^2 = \frac{1}{m}\sum_{i=1}^{m} X_i^2, \quad \hat{\sigma}_2^2 = s_2^2 = \frac{1}{n}\sum_{i=1}^{n} Y_i^2$$

Under the hypothesis $H_0$, the statistic $F$ has the Fisher distribution with $m$ and $n$ degrees of freedom. The hypothesis is rejected with a significance level $\alpha$ if $F > F_\alpha(m, n)$.

We have $s_1^2 = 3.2024$, $s_2^2 = 0.8978$, $F = 3.567$ and the $P$-value is

$$pv = \mathbf{P}\{F_{m,n} > 3.567\} = 0.0075$$

In this example, the parametric Fisher test gives almost the same $P$-value as the general Siegel–Tukey or Ansari–Bradley tests, and reject the hypothesis very evidently. The Klotz and Mood tests are not so powerful.

## 4.6. Hypothesis on median value: the Wilcoxon signed ranks test

Suppose that $X_1, ..., X_n$ is a simple sample of the random variable $X$ with the cdf $F$ belonging to the class $\mathcal{F}$ of absolutely continuous cumulative distribution functions with finite first two moments.

Denote by $M$ the median of random variable $X$ and let us consider the following hypothesis.

**Hypothesis on median value:**

$$H_0 : F \in \mathcal{F}, \; M = M_0;$$

where $M_0$ is a fixed value of the median.

**One-sided alternatives:** $H_1 : F \in \mathcal{F}, \; M > M_0$ and $H_2 : F \in \mathcal{F}, \; M < M_0$. **Two-sided alternative:** $H_3 : F \in \mathcal{F}, \; M \neq M_0$.

### 4.6.1. Wilcoxon's signed ranks tests

The Wilcoxon's signed ranks test is used when the class $\mathcal{F}$ consists of symmetric distributions.

If the class of alternatives includes non-symmetric distributions then the *sign test* (see section 5.1) may be used.

If the class $\mathcal{F}$ consists of symmetric distributions and $n$ is large, a powerful competitor to the Wilcoxon's signed ranks test is the asymptotic Student's test. This test is used to test the hypothesis $\tilde{H}_0 : F \in \mathcal{F}$, $\mu = \mu_0$, where $\mu = \mathbf{E}X_i$ is the mean, but, in the case of symmetric distributions, the median coincides with the mean: $M = \mu$, so the hypothesis $\tilde{H}_0$ is equivalent to the hypothesis $H_0$. The Student's asymptotic test is based on the statistic

$$t = \sqrt{n}\,\frac{\bar{X} - \mu_0}{s}$$

where

$$\bar{X} = \frac{1}{n}\sum_{i=1}^{n} X_i, \quad s^2 = \frac{1}{n-1}\sum_{i=1}^{n}(X_i - \bar{X})^2$$

Under the hypothesis $\tilde{H}_0$

$$\sqrt{n}(\bar{X} - \mu_0)/\sigma \overset{d}{\to} Z \sim N(0,\,1), \quad s \overset{P}{\to} \sigma, \quad t \overset{d}{\to} Z \sim N(0,\,1)$$

where $\sigma^2 = \mathbf{Var}X_i$. So the following test is used against two-sided alternatives.

**Asymptotic Student's test:** if $n$ is large and the alternative is two-sided then the hypothesis $\tilde{H}$ is rejected with an asymptotic significance level $\alpha$ if $|t| > z_0\alpha/2$.

**Construction of Wilcoxon's signed ranks test**

Set $D_i = X_i - M_0$ and denote by $R_i$ the rank of the element $|D_i|$ in the sequence $|D|_1, \ldots, |D|_n$, $T^+$ and by $T^-$ the sums of the ranks corresponding to the positive and negative differences $D_i$

$$T^+ = \sum_{i:D_i>0} R_i, \quad T^- = \sum_{i:D_i<0} R_i \qquad [4.57]$$

For example, if $M_0 = 10$ and the realization of the sample is 7, 16, 5, 8, 14 then the realization of $D_1, \ldots, D_5$

is $-3$, $6$, $-5$, $-2$, $4$, of $|D|_1, \ldots,$ $|D|_n$ is   $3$, $6$, $5$, $2$, $4$, and of $R_1, \ldots, R_5$ is   $2$, $5$, $4$, $1$, $3$. So, the rank values corresponding to positive and negative $D_i$ values are $3, 5$ and $1, 2, 4$, respectively, and $T^+ = 3 + 5 = 8$, $T^- = 1 + 2 + 4 = 7$.

In practice, $T^+$ and $T^-$ are found as follows: the realization $-3$, $6$, $-5$, $-2$, $4$ of $D_1, \ldots, D_5$ is arranged in order of increasing absolute values but keeping the signs: $-2$, $-3$, $4$, $-5$, $6$. The "signed ranks" are $-1$, $-2$, $3$, $-4$, $5$. So, $T^+ = 3 + 5 = 8$, $T^- = 1 + 2 + 4 = 7$.

It is sufficient to use one of the statistics $T^+$ or $T^-$ because they are related by the equality $T^+ + T^- = R_1 + \cdots + R_n = n(n+1)/2$. Possible values of the statistic $T^+$ are $0, 1, \ldots, n(n+1)/2$.

Under the hypothesis $H_0$, the distribution of the difference $D_i$ is symmetric with respect to zero, so the distributions of sums of the ranks $T^+$ and $T^-$ coincide and $\mathbf{E}T^+ = \mathbf{E}T^-$.

If $M > M_0$ then the distribution of the random variable $D_i$ is symmetric with respect to $\theta = M - M_0 > 0$, so for any $c > 0$ the random variable $D_i$ takes values in the intervals $(\theta - c, \theta)$ and $(\theta, \theta + c)$ with equal probabilities, hence $D_i$ tend to take positive rather than negative values. Moreover, if the positive and negative values are equidistant from $\theta$ then the absolute value of the negative value is smaller than the absolute value of the positive value. So the values of $T^+$ tend to be greater than those of $T^-$, and $\mathbf{E}T^+ > \mathbf{E}T^-$.

Analogously, if $M < M_0$ then the values of $T^+$ tend to be smaller than those of $T^-$, and $\mathbf{E}T^+ < \mathbf{E}T^-$.

Since the sum of all ranks $T^+ + T^- = n(n+1)/2$ is constant, the conditions

$$\mathbf{E}T^+ = \mathbf{E}T^-, \quad \mathbf{E}T^+ > \mathbf{E}T^-, \quad \mathbf{E}T^+ < \mathbf{E}T^-$$

are equivalent to the conditions

$$\mathbf{E}T^+ = n(n+1)/4 =: N, \quad \mathbf{E}T^+ > N, \quad \mathbf{E}T^+ < N$$

So if $M = M_0$ then the values of the statistic $T^+$ are spread around $N$; if $M > M_0$ they are spread around a number greater than $N$; if $M < M_0$ they are spread around a number smaller than $N$. We obtain the following test.

**Wilcoxon's signed ranks test:** in the case of a two-sided alternative $H_3$, the hypothesis $H_0$ is rejected with a significance level not greater than $\alpha$ if

$$T^+ \geq T^+_{\alpha/2}(n) \quad \text{or} \quad T^+ \leq T^+_{1-\alpha/2}(n) \qquad [4.58]$$

where $T^+_\alpha(n)$ is the smallest real number verifying the inequality $\mathbf{P}\{T^+ \geq T^+_{\alpha/2}(n)\} \leq \alpha/2$ and $T^+_{1-\alpha/2}$ is the largest real number verifying the inequality $\mathbf{P}\{T^+ \leq T^+_{1-\alpha/2}\} \leq \alpha/2$. In the case of the one-sided alternatives $H_1$ and $H_2$, the hypothesis $H$ is rejected if

$$T^+ \geq T^+_\alpha(n) \quad \text{or} \quad T^+ \leq T^+_{1-\alpha} \qquad [4.59]$$

respectively.

**Comment 4.12.**    If $\mathcal{F}$ contains non-symmetric probability densities then Wilcoxon's signed ranks test can be inefficient.

Suppose that the alternative is $H_1 : F \in \mathcal{F}$, $M > M_0$; $\theta = M - M_0 > 0$ is the median of $D_i$. Suppose that the pdf of the random variable $D_i$ has a light and long left "tail" and a short and heavy right "tail". Since $\mathbf{P}\{D_i > \theta\} = \mathbf{P}\{D_i < \theta\}$, as in the symmetric case, $D_i$ tend to take positive rather than negative values. But, unlike the symmetric case, the extreme negative $D_i$ tend to take larger absolute values than the do the extreme positive $D_i$, so the sum of ranks $T^+$ can be near to the sum $T^-$, or even smaller. So the test will not reject the alternative.

If $n$ is not large ($n \leq 30$), the critical values $T_\alpha^+(n)$ (and $P$-values) are found using the probability distribution of the statistic $T^+$ given in the following theorem.

**Theorem 4.10.** *If the hypothesis $H_0$ holds then*

$$\mathbf{E}(T^+) = \frac{n(n+1)}{4}, \quad \mathrm{Var}(T_+) = \frac{n(n+1)(2n+1)}{24}$$

$$\mathbf{P}\{T^+ = k\} = \frac{c_{kn}}{2^n}, \quad k = 0, 1, \ldots, n(n+1)/2 \qquad [4.60]$$

*where $c_{kn}$ are coefficients of $t^k$ in the product $\prod_{k=1}^{n}(1 + t^k)$.*

**Proof.** The statistic $T^+$ can be written in another form. Let us consider the ordered sequence $|D|_{(1)}, \ldots, |D|_{(n)}$ obtained from $|D_1|, \ldots, |D_n|$. Set

$$W_i = \begin{cases} 1, & \text{if } \exists D_j > 0 : |D|_{(i)} = D_j \\ 0, & \text{otherwise.} \end{cases}$$

So, $W_i = 1$ if $D_j > 0$: $R_j = i$, hence

$$T^+ = \sum_{i=1}^{n} i W_i \qquad [4.61]$$

Since the random variables $D_1, \ldots, D_n$ are independent, take positive and negative values with the equal probability 0.5, so for all $k_1, \ldots, k_n \in \{0, 1\}$

$$\mathbf{P}\{W_1 = k_1, \ldots, W_n = k_n\} = \left(\frac{1}{2}\right)^n, \quad \mathbf{P}\{W_i = k_i\} = \frac{1}{2}$$

so $W_1, \ldots, W_n$ are i.i.d. Bernoulli random variables: $W_i \sim B(1, 1/2)$. Hence

$$\mathbf{E}(T^+) = \sum_{i=1}^{n} i \frac{1}{2} = \frac{n(n+1)}{4}, \quad \mathrm{Var}(T_+) = \sum_{i=1}^{n} i^2 \frac{1}{4} = \frac{n(n+1)(2n+1)}{24}$$

The generating function

$$\psi(t) = \mathbf{E}(t^{T^+}) = \sum_{k=0}^{M} t^k \mathbf{P}\{T^+ = k\} \qquad [4.62]$$

of the random variable $T^+$ has the form

$$\psi(t) = \prod_{i=1}^{n} \mathbf{E} t^{iW_i} = \frac{1}{2^n} \prod_{i=1}^{n} (1 + t^i) = \sum_{k=0}^{M} c_{kn} t^k \qquad [4.63]$$

[4.62] and [4.63] imply [4.60].

$\triangle$

**Example 4.8.** In a stock exchange the returns (euros per share) are:

$$3.45; 4.21; 2.56; 6.54; 3.25; 7.11.$$

Test the hypotheses: a) the median return is not larger than 4 euros; b) the median return is greater than 4 euros; c) the median return is not greater than 3 euros; d) the median return is greater than 7 euros; and e) the median return is equal to 3 euros. The significance level is $\alpha = 0.1$.

Since $M_0 = 4$, find the differences $D_i = X_i - 4$: $-0.55$; $0.21$; $-1,44$; $2.54$; $-0.75$; $3.11$. Order them keeping the signs: $0.21$; $-0.55$; $-0.75$; $-1,44$; $2.54$; $3.11$. The "signed" ranks are 1; $-2$; $-3$; $-4$; 5; 6. There are three positive and three negative differences. The sum of ranks $T^+$ takes the value $t^+ = 1 + 5 + 6 = 12$. In case a) the $P$-value is

$$pv = \mathbf{P}\{T^+ \geq 12|H_0\} = 0.422$$

In case b) the $P$-value is

$$pv = \mathbf{P}\{T^+ \leq 12|H - 0\} = 0.656$$

The data do not contradict the first two hypotheses.

c) Find the differences $D_i = X_i - 3$: 0.45; 1.21; $-0.44$; 3.54; 0.25; 4.11. Order them: 0.25; $-0.44$; 0.45; 1.21; 3.54; 4.11. The "signed" ranks are 1; $-2$; 3; 4; 5; 6, $t^+ = 19$. So

$$pv = \mathbf{P}\{T^+ \geq 19|H_0\} = 0.047$$

The hypothesis is rejected.

d) Find the differences $D_i = X_i - 7$: $-3.55$; $-2.79$; $-4.44$; $-0.46$; $-3.75$; 0.11. There is only one positive difference and its "signed" rank is 1, so $t^+ = 1$. The $P$-value is

$$pv = \mathbf{P}\{T^+ \leq 1|H_0\} = 0.031$$

The hypothesis is rejected.

e) In c) we obtained $t^+ = 19$. So

$$pv = 2\min(\mathbf{P}\{T^+ \leq 19|H\}$$

$$\mathbf{P}\{T^+ \geq 19|H\}) = 2\min(0.031; 0.047) = 0.062$$

The hypothesis is rejected.

**Comment 4.13.** If some differences $D_i$ are equal to zero (as a result of rounding) then they are excluded and the test is conditional, given the number of remaining differences.

**Large samples.** In the case of large samples an approximate test is obtained using the limit distribution of the statistic $T^+$.

**Theorem 4.11.** *If the hypothesis $H_0$ holds then*

$$Z_n = \frac{T^+ - \mathbf{E}(T^+)}{\sqrt{\mathbf{Var}(T_+)}} \xrightarrow{d} Z \sim N(0, 1)$$

**Proof.** Since

$$\frac{T^+ - \mathbf{E}(T^+)}{\sqrt{\mathbf{Var}(T_+)}} = \sum_{i=1}^{n} Y_i, \quad Y_i = \frac{iW_i - i/2}{\sqrt{\mathbf{Var}(T_+)}}$$

$$\mathbf{E}Y_i = 0, \quad \mathbf{Var}(\sum_{i=1}^{n} Y_i) = 1$$

$$\mathbf{E}|Y_i|^3 = \mathbf{E}\left|\frac{|iW_i - i/2|}{\mathbf{Var}(T_+)}\right|^3 = \frac{(i/2)^3}{[n(n+1)(2n+1)/24]^{3/2}} \leq \frac{n^3}{8[2n^3/24]^{3/2}}$$

The result of the theorem is implied by Liapunov's theorem.

$\triangle$

**Asymptotic Wilcoxon's signed ranks test:** If $n$ is large then the hypothesis $H_0$ is rejected with an approximate significance level $\alpha$ if

$$|Z_n| \geq z_{\alpha/2}$$

If coinciding ranks are present then the statistic is modified

$$Z_n^* = \frac{Z_n}{\sqrt{1 - T/(2n(n+1)(2n+1))}}$$

where $T = \sum_{l=1}^{k}(t_l^3 - t_l)$, $k$ is the number of coinciding rank groups, $t_l$ is the size of the $l$-th group. The hypothesis $H_0$ is rejected with an approximate significance level $\alpha$ if

$$|Z_n^*| \geq z_{\alpha/2}$$

**Example 4.9.** (continuation of Example 3.2) Using the data in Example 3.2, test the hypothesis that the median of the random variable $V$ is a) greater than 15; b) equal to 15.

The number of positive differences $D_i = X_i - 15$ is 15, $T^+ = 347.5$. The statistic

$$Z_n = \frac{T^+ - \frac{n(n+1)}{4}}{\sqrt{\frac{n(n+1)(2n+1)}{24}}} = \frac{347,5 - \frac{49(49+1)}{4}}{\sqrt{\frac{49(49+1)(249+1)}{24}}} = -2.63603$$

There are three groups with two coinciding ranks: $(12.5; 12.5)$, $(14.5; 14.5)$, $(33.5; 33.5)$ and one group with four coinciding ranks: $(18.5; 18.5; 18.5; 18.5)$. So $k = 4$, $t_1 = t_2 = t_3 = 2$, $t_4 = 4$ and $T = 3(2^3 - 2) + (4^3 - 4) = 78$. The correction is very small

$$\sqrt{1 - T/(2n(n+1)(2n+1))} = 0.999920$$

so    $Z_n^* = \dfrac{Z_n}{0.999920} = -2.63677$

a) The hypothesis is rejected if the statistic $Z_n^*$ takes small values. The asymptotic $P$-value is

$$pv_a = \Phi(-2.63677) = 0.00419$$

The hypothesis is rejected. The exact $P$-value computed using the SPSS package is $pv = 0.003815$.

b) The hypothesis is rejected if the statistic $|Z_n^*|$ takes large values. The asymptotic $P$-value is

$$pv_a = 2(1 - \Phi(2.63677)) = 0.00838$$

The exact $P$-value computed using the SPSS package is $pv = 0.00763$.

**Comment 4.14.**   Some statistical packages use the following modification: instead of statistic $T^+$ the statistic

$$T = \min\left(T^+, T^-\right) - \frac{n(n+1)}{4} = \min\left(T^+, \frac{n(n+1)}{2} - T^+\right) - \frac{n(n+1)}{4}$$
$$[4.64]$$

is used. This statistic takes non-positive values and under hypothesis $H_0$ most of its values are near $0$. The distribution of this statistic is chosen using the distribution of the statistics $T^+$. The hypothesis $H_0$ is rejected if $T < T_{1-\alpha}$, where $T_{1-\alpha}$ is the $\alpha$-critical level of the statistics $T$. Theorem 4.11 implies that for all $x < 0$

$$\mathbf{P}\{\frac{T}{\sqrt{\text{Var}(T_+)}} \le x\} \to 2\Phi(x)$$

so for large $n$, the hypothesis $H_0$ is rejected if

$$T/\sqrt{\mathbf{Var}(T_+)} < -z_\alpha$$

where $\Phi(x)$ and $z_\alpha$ are the cumulative distribution function and $\alpha$-critical value of the standard normal distribution.

### 4.6.2. ARE of the Wilcoxon signed ranks test with respect to Student's test

Let us find the ARE of the Wilcoxon's signed ranks test for the median with respect to the Student's test for the mean when the distribution is symmetric (in such a case the median coincides with the mean and the same hypothesis is tested).

Denote by $f_D(x)$ the pdf $F_D(x)$ – the cdf of the random variable $D$ under the hypothesis $H_0$. The pdf $f_D(z)$ is symmetric with respect to zero: $f_D(-x) = f_D(x)$.

Under the alternative $H_3$ the pdf of the random variable $D$ is $f_D(x|\theta) = f_D(x - \theta)$, $\theta = M - M_0$, and the cdf is $F_D(x|\theta) = F_D(x - \theta)$.

**Theorem 4.12.** *The ARE of the Wilcoxon's signed ranks test with respect to the Student's test for related samples is*

$$e(T^+, t) = 12\tau^2 \left[\int_{-\infty}^{\infty} f_D^2(x)\, dx\right]^2, \quad \tau^2 = \mathbf{Var}(D_i) \qquad [4.65]$$

**Proof.** The statistics $T^+$ can be written in the form

$$T^+ = \sum_{1 \le i \le j \le n} T_{ij}, \quad T_{ij} = \begin{cases} 1, & \text{if } D_i + D_j > 0 \\ 0, & \text{otherwise.} \end{cases} \qquad [4.66]$$

Let us consider the ordered sample $D_{(1)} \le \dots \le D_{(n)}$. Suppose that $D_{(k)} < 0$, $D_{(k+1)} > 0$. We obtain

$$\sum_{1 \le i \le j \le n} \mathbf{1}_{\{D_i + D_j > 0\}} = \sum_{1 \le i \le j \le n} \mathbf{1}_{\{D_{(i)} + D_{(j)} > 0\}} =$$

$$\sum_{j=k+1}^{n} \sum_{i=1}^{j} 1_{\{D_{(i)}+D_{(j)}>0\}} = \sum_{j=k+1}^{n} \left( \sum_{i=1}^{k} 1_{\{D_{(i)}+D_{(j)}>0\}} + \sum_{i=k+1}^{j} 1 \right) =$$

$$\sum_{j=k+1}^{n} \left( \sum_{i=1}^{k} 1_{\{|D_{(j)}|>|D_{(i)}|\}} + j - k \right) = T^+$$

Under the alternative

$$\mathbf{E}_\theta T^+ = n\mathbf{E}_\theta T_{11} + \frac{n(n-1)}{2} \mathbf{E}_\theta T_{12}$$

$$\mathbf{E}_\theta T_{11} = \mathbf{P}_\theta\{D_1 > 0\} = 1 - F_D(-\theta), \quad \mathbf{E}_\theta T_{12} = \mathbf{P}_\theta\{D_1 + D_2 > 0\} =$$

$$\int_{-\infty}^{\infty} \mathbf{P}_\theta\{D_1 > -x\} dF_{D_2}(x|\theta) = \int_{-\infty}^{\infty} [1 - F_D(-y - 2\theta)] dF_D(y)$$

Since the test based on the statistic $T^+$ is equivalent to the test based on $V^+ = T^+/C_n^2$, $C_n^2 = n(n-1)/2$, let us consider the statistic $V^+$. We obtain

$$\mu_{1n}(\theta) = \mathbf{E}_\theta V^+ = \frac{2}{n-1}[1 - F_D(-\theta)] + \int_{-\infty}^{\infty} [1 - F_D(-y - 2\theta)] dF_D(y) \rightarrow$$

$$\int_{-\infty}^{\infty} [1 - F_D(-y - 2\theta)] dF_D(y) = \mu_1(\theta)$$

$$\dot{\mu}_1(\theta) = 2 \int_{-\infty}^{\infty} f_D(-y - 2\theta) dF_D(y)$$

Using the symmetry of $f_D$ we have

$$\dot{\mu}_1(0) = 2 \int_{-\infty}^{\infty} f_D^2(y) dy$$

Under the hypothesis, the variance of the random variable $T^+$ is given by Theorem 4.10

$$\mathbf{Var}_0(T_+) = n(n+1)(2n+1)/24$$

so

$$\sigma_{1n}^2(0) = \mathbf{Var}_0(V^+) = (n+1)(2n+1)/(6n(n^2-1))$$

hence

$$n\sigma_{1n}^2(0) \to \sigma_1^2(0) = \frac{1}{3}$$

The Student's test for related samples is based on the statistic $t = \sqrt{n}\bar{D}/s$, which is asymptotically equivalent to the statistic $\bar{D}$ because $s$ tends to a constant in probability.

We have

$$\mu_2(\theta) = \mathbf{E}_\theta \bar{D} = \theta, \quad \dot{\mu}_2(\theta) = 1$$

$$\sigma_{2,n}^2(0) = \mathbf{Var}_0 \bar{Z} = \tau^2/n, \quad \tau^2 = \mathbf{Var}_0 D_i, \quad n\sigma_{2n}^2(0) \to \tau^2 = \sigma_2^2(0)$$

The ARE of the Wilcoxon's signed ranks test with respect to the Student's test is

$$e(T^+, t) = \left( \frac{\dot{\mu}_1(0)\sigma_2(0)}{\sigma_1(0)\dot{\mu}_2(0)} \right)^2 = \left( \frac{2 \int_{-\infty}^{\infty} f_D^2(x)\, dx\, \tau}{\frac{1}{\sqrt{3}} \cdot 1} \right)^2 =$$

$$12\tau^2 \left[ \int_{-\infty}^{\infty} f_D^2(x)\, dx \right]^2$$

$\triangle$

**Comment 4.15.** The ARE is computed using the same formula as the ARE of the Wilcoxon–Mann–Whitney test with respect to the Student's test in the case of independent samples. So, the Wilcoxon's signed ranks test is good "competition" for the Student's test. Its ARE is near 1 if the distribution of $D_i$ is normal (ARE=$3/\pi \approx 0.955$), equal to 1 in the case of a uniform distribution, and even greater than one if the distribution of $D_i$ has heavy tails as, for example, in the case of logistic (ARE = $\pi^2/9 \approx 1.097$), extreme values (ARE=$\pi^2/8 \approx 1.097$), and doubly exponential (ARE = 1,5) distributions. Theorem 4.9 implies that the ARE cannot be smaller than 0.864.

### 4.7. Wilcoxon's signed ranks test for homogeneity of two related samples

Suppose that

$$(X_1, Y_1)^T, \ldots, (X_n, Y_n)^T$$

is a simple sample of the random vector $(X, Y)^T$ having the cdf $F(x, y)$ from the class $\mathcal{F}$ of absolutely continuous cdf with finite first two moments. Denote by $F_1(x)$ and $F_2(y)$ the marginal cdf.

#### Homogeneity hypothesis of two related samples

$$H_0 : F \in \mathcal{F}, \ F_1(x) \equiv F_2(x)$$

The two-sample Wilcoxon's signed ranks test is used when the cdf of the class $\mathcal{F}$ have the form

$$F(x, y) = F_0(x, y + \theta), \quad \theta \in \mathbf{R}$$

where $F_0(x, y)$ is a symmetric cdf, so $(X, Y - \theta)^T$ has a symmetric distribution.

If the class of alternatives includes non-symmetric distributions of $(X, Y - \theta)^T$ then the hypothesis $H_0$ can be verified using the *sign test*. This will be discussed in section 5.1.

Set $D = X - Y$. Denote by $M$ the median of the random variable $D$ and let us consider the following hypothesis.

#### Hypothesis on the zero value of the median of $D$

$$H_0^* : F \in \mathcal{F}, \ M = 0$$

This hypothesis is wider then the homogeneity hypothesis $H_0$ because under $H_0$ the cdf $F(x, y)$ is symmetric, so $\mathbf{P}\{D > 0\} = \mathbf{P}\{D < 0\}$, which means that $M = 0$.

The hypothesis $H_0^*$ can be tested using the sample $D_1, \ldots, D_n$ of the differences $D_i = X_i - Y_i$; the one-sample Wilcoxon's signed ranks test for the median (and, in the case of large samples, the Student's asymptotic test) can be used. All the results from the previous section can be applied.

The *two-sample Wilcoxon's signed ranks statistic* has the same form and the same distribution as the one-sample statistic considered in section 4.5, replacing the differences $X_i - M_0$ by the differences $D_i = X_i - Y_i$ and $M_0$ by $0$.

If the hypothesis on the median value $H_0^*$ is rejected then the homogeneity hypothesis of two related samples $H_0$ is also rejected because it is narrower.

**Example 4.10.** (continuation of Example 4.1). Test the hypothesis $H_0$, that the median of the difference $D = X - Y$ of coordinates given in exercise 4.1 is equal to 0.

We find the differences $D_i = X_i - Y_i, i = 1, \ldots, 50$, and compute $T^+ = 718.5$. Under $H_0$ (see Theorem 4.10), we get $\mathbf{E}T^+ = 637.5, \mathbf{Var}(T^+) = 10731.25$. We compute $Z_n = 0.7819$. There are some *ex aequo* among $|D_i|$, so we compute the value of the modified statistic $Z_n^* = 0.7828$. The asymptotic $P$-value $pv_a = 2(1 - \Phi(0.7828)) = 0.4337$. The data do not contradict the hypothesis.

## 4.8. Test for homogeneity of several independent samples: Kruskal–Wallis test

Suppose that

$$\mathbf{X}_1 = (X_{11}, \ldots, X_{1n_1})^T, \quad \ldots \quad, \quad \mathbf{X}_k = (X_{k1}, \ldots, X_{kn_k})^T$$

are $k$ simple samples of independent random variables $X_1, \ldots, X_k$ with absolutely continuous cumulative distribution functions $F_1(x), \ldots, F_k(x)$.

**Homogeneity hypothesis for $k > 2$ independent samples:**

$$H_0 : F_1(x) = F_2(x) = \cdots = F_k(x) =: F(x), \ \forall x \in \mathbf{R} \qquad [4.67]$$

Suppose that the location alternative is considered

$$H_1 : F_j(x) = F(x - \theta_j) \ \text{for all } x \in \mathbf{R}, \quad j = 1, \ldots, k, \quad \sum_{j=1}^{k} \theta_j^2 > 0$$

$$[4.68]$$

If all distributions $F_j$ are normal with the same variance $\sigma^2$, and possibly different means $\mu_j$, then the alternative $H_1$ is

$$H_1 : \mu_j = \mu + \theta_j, \ \text{for all } x \in \mathbf{R}, \quad j = 1, \ldots, k, \quad \sum_{j=1}^{k} \theta_j^2 > 0$$

and the hypothesis $H_0$ is parametric

$$H_0 : \mu_1 = \ldots = \mu_k$$

A well-known test for the equality of means of normal distributions when variances are constant is considered in the one-factor analysis of the variance and is based on statistics

$$F = \frac{1}{k-1} \sum_{i=1}^{k} n_i (\bar{X}_{i.} - \bar{X}_{..})^2 \Big/ \frac{1}{n-k} \sum_{i=1}^{k} \sum_{j=1}^{n_i} n_i (X_{ij} - \bar{X}_{i.})^2 \quad [4.69]$$

where $\bar{X}_{i.}$ is the arithmetic mean of the $i$-th sample, $\bar{X}_{..}$ is the arithmetic mean of the unified sample, $n = \sum_{i=1}^{k} n_i$.

If hypothesis [4.67] holds in the normal case then statistic [4.69] has a Fisher distribution with $k - 1$ and $n - k$ degrees of freedom. If $n_i \to \infty$, $n_i/n \to p_i \in (0, 1)$, $k$ is fixed then the denominator of [(4.69] converges to $\sigma^2$, and the distribution of the statistic $F$ converges to the distribution of the random variable $\chi^2_{k-1}/(k-1)$.

If $n_i$ are large then the statistic $F$ is proportional to the numerator.

**Idea of the test.** The observations $X_{i1}, ..., X_{in_i}$ are replaced by their ranks $R_{i1}, ..., R_{in_i}$ in the unified ordered sample of all observations.

The *Kruskal–Wallis* statistic is defined in a similar way to the Fisher statistic [4.69], replacing $\bar{X}_{i.}$ by $\bar{R}_{i.}$ and $\bar{X}_{..}$ by $\bar{R}_{..} = (n+1)/2$ in the numerator and multiplying the obtained expression by the proportionality coefficient

$$F_{KW} = \frac{12}{n(n+1)} \sum_{i=1}^{k} n_i \left( \bar{R}_{i.} - \frac{n+1}{2} \right)^2 = \frac{12}{n(n+1)} \sum_{i=1}^{k} n_i \bar{R}_{i.}^2 - 3(n+1)$$

$$= \frac{12}{n(n+1)} \sum_{i=1}^{k} \frac{R_{i.}^2}{n_i} - 3(n+1) \qquad [4.70]$$

where

$$R_{i.} = \sum_{j=1}^{n_i} R_{ij}$$

is the sum of ranks of the elements of the $i$-th sample in the unified sample $i = 1, \ldots, k$.

The proportionality coefficient is chosen to obtain the limit of the chi-squared distribution of the statistics $F_{KW}$ under the hypothesis $H_0$, as sample sizes tend to infinity.

Under the hypothesis $H_0$ :

$$\mathbf{E}(\bar{R}_{i.}) = \mathbf{E}(R_{ij}) = (n+1)/2$$

for all $i$, under the alternative, some means are greater or smaller than $(n+1)/2$, so the statistic $F_{KW}$, which is based on the sums of the terms $(\bar{R}_{i.} - (n+1)/2)^2$, tends to take greater values under the alternative.

**Kruskal–Wallis test:** the hypothesis $H_0$ is rejected if $F_{KW} > F_{KW}(\alpha)$, where $F_{KW}(\alpha)$ is the minimum real number $c$ verifying the inequality $\mathbf{P}\{F_{KW} > c|H_0\}$.

**Large samples.** Let us prove that if the sample sizes are large then the probability distribution of the statistic $F_{KW}$ can be approximated by the chi-squared distribution with $k - 1$ degrees of freedom.

**Theorem 4.13.** *If the hypothesis $H_0$ holds, $n_i/n \to p_{i0} \in (0, 1)$, $i = 1, ..., k$, then*

$$F_{KW} \xrightarrow{d} S \sim \chi^2(k - 1) \qquad [4.71]$$

**Proof.** Denote by $R_i = \sum_{j=1}^{n_i} R_{ij}$ the sum of the ranks of the $i$-th group in the unified sample.

Since the variance in $R_i$ coincides with the variance in Wilcoxon's rank sum statistic when the $i$-th group is interpreted as the first, and all others as the second, we obtain

$$\mathbf{E}R_i = n_i \frac{n + 1}{2}, \quad \mathbf{Var}(R_i) = \frac{n_i(n + 1)(n - n_i)}{12}$$

Moreover, for all $i \neq j$

$$\mathbf{Cov}(R_i, R_j) = \sum_{l=1}^{n_i} \sum_{s=1}^{n_j} \mathbf{Cov}(R_{il}, R_{js}) = -n_i n_j \frac{n + 1}{12}$$

So the covariance matrix of the random vector $\mathbf{R} = (R_1, ..., R_k)^T$ is $\mathbf{\Sigma}_n = [\sigma_{ij}]_{k \times k}$, where

$$\sigma_{ij} = \begin{cases} n_i(n + 1)(n - n_i)/12, & i = j \\ -n_i n_j(n + 1)/12, & i \neq j \end{cases}$$

We obtain

$$\frac{12}{(n + 1)n^2} \mathbf{\Sigma}_n =$$

$$\begin{pmatrix} p_1(1-p_1) & -p_1p_2 & \cdots & -p_1p_k \\ -p_2p_1 & p_2(1-p_2) & \cdots & -p_2p_k \\ \cdots & \cdots & \cdots & \cdots \\ -p_kp_1 & -p_kp_2 & \cdots & p(1-p_k) \end{pmatrix}$$

$$= D - pp^T$$

where $p_i = n_i/n$, $p = (p_1,...,p_k)^T$, $D$ is the diagonal matrix with the diagonal elements $p_1,...,p_k$.

The generalized inverse of the matrix $A = D - pp^T$ is

$$A^- = (D - pp^T)^- = D^{-1} + \frac{1}{p_k}\mathbf{1}\mathbf{1}^T, \quad \mathbf{1} = (1,...,1)^T$$

Using the equalities

$$\mathbf{1}^T D = p^T, \quad \mathbf{1}^T p = p^T \mathbf{1} = 1, \quad D\mathbf{1} = p, \quad p^T D^{-1} = \mathbf{1}^T$$

we obtain. $AA^-A = A$ So

$$\Sigma_n^- = \frac{12}{n^2(n+1)}[\sigma^{ij}]_{k \times k}, \quad \sigma^{ii} = \frac{1}{p_k} + \frac{1}{p_i}, \quad \sigma^{ij} = \frac{1}{p_k}, \ i \neq j$$

We have

$$(R - ER)^T \Sigma_n^- (R - ER) =$$

$$\frac{12}{n(n+1)} \sum_{i=1}^{k} \frac{1}{n_i}\left(R_i - \frac{n_i(n+1)}{2}\right)^2 = F_{KW}$$

Note that

$$n^{-3}\Sigma_n \to \Sigma = \frac{1}{12}(D_0 - p_0 p_0^T)$$

where $p_0 = (p_{10},...,p_{k0})^T$, and $D_0$ is diagonal with diagonal elements $p_{10},...,p_{k0}$. The rank of the matrix $\Sigma$ is $k-1$ (see exercise 2.4).

By the CLT for sequences of random vectors

$$n^{-3/2}(R - E(R)) \xrightarrow{d} Z \sim N_k(0, \Sigma)$$

where

$$\Sigma = \frac{1}{12}(D_0 - p_0 p_0^T), \quad p_0 = (p_{10}, \ldots, p_{k0})^T$$

and $D_0$ is the diagonal matrix with diagonal elements $p_{10}, \ldots, p_{k0}$.

We shall use a theorem stating that if

$$X \sim N_k(\mu, \Sigma) \quad \text{then} \quad (X - \mu)^T \Sigma^-(X - \mu) \sim \chi^2(r)$$

where $\Sigma^-$ is a generalized inverse of the matrix $\Sigma$, $r$ is the rank of the matrix $\Sigma$.

By this theorem

$$n^{-3}(R - E(R))^T \Sigma^-(R - E(R)) \xrightarrow{d} S \sim \chi^2(k-1)$$

hence the result of theorem.

$\triangle$

**Asymptotic Kruskal–Wallis test:** if $n_i$ are large then the hypothesis $H_0$ is rejected with an approximate significance level $\alpha$ if $F_{KW} > \chi_\alpha^2(k-1)$.

**ARE of the Kruskal–Wallis test with respect to Fisher's test.** If $n \to \infty$, $k$ is fixed, then the ARE of the Kruskal–Wallis test with respect to Fisher's test is

$$e(KW, F) = 12\sigma^2 \left[ \int_{-\infty}^{\infty} f^2(x)dx \right]^2$$

where $f$ and $\sigma^2$ are the probability density and the variance of the random variable $X_{ij}$ under the zero hypothesis.

The formula is identical to the formula for the ARE of the Wilcoxon's rank sum test with respect to the Student's test.

The ARE is near to 1 if the distribution of $D_i$ is normal (ARE = $3/\pi \approx 0.955$), equal to 1 under a uniform distribution, and is greater than 1 if this distribution has heavy tails as, for example, in the case of logistic (ARE = $\pi^2/9 \approx 1.097$), extreme values (ARE = $\pi^2/8 \approx 1.234$) and doubly exponential (ARE = 1,5) distributions.

Theorem 4.9 implies that the ARE cannot be smaller than 0.864.

**Data with** *ex aequo*. If there are coinciding values then, as for the Wilcoxon's rank sum test, the statistic $F_{KW}$ is modified

$$F^*_{KW} = F_{KW}/(1 - T/(n^3 - n))$$

where $T = \sum_{i=1}^{s} T_i$, $T_i = (t_i^3 - t_i)$, $s$ is the number of coinciding observations groups in the unified sample, $t_i$ is the size of the $i$-th group.

**Modified asymptotic Kruskal–Wallis test:** the hypothesis $H_0$ is rejected with an approximate significance level $\alpha$ if $F^*_{KW} > \chi^2_\alpha(k - 1)$.

**Example 4.11.** The quantity of serotonin is measured after injections of three pharmaceuticals. The serotonin quantities of patients in three groups are given below. Have the three medications equal influence on the quantity of serotonin?

| 1 (Placebo) | 2 (remedy 1) | 3 (remedy2) |
|:---:|:---:|:---:|
| 340 | 294 | 263 |
| 340 | 325 | 309 |
| 356 | 325 | 340 |
| 386 | 340 | 356 |
| 386 | 356 | 371 |
| 402 | 371 | 371 |
| 402 | 385 | 402 |
| 417 | 402 | 417 |
| 433 | | |
| 495 | | |
| 557 | | |

The unified ordered sample is (the group number is in parentheses):

| 1 | 2 | 3 | 4 | 5 | 6 | 7 | 8 |
|:---:|:---:|:---:|:---:|:---:|:---:|:---:|:---:|
| 263(3) | 294(2) | 309(3) | 325(2) | 325(2) | 340(1) | 340(1) | 340(2) |

| 9 | 10 | 11 | 12 | 13 | 14 | 15 | 16 |
|:---:|:---:|:---:|:---:|:---:|:---:|:---:|:---:|
| 340(3) | 356(1) | 356(2) | 356(3) | 371(2) | 371(3) | 371(3) | 385(2) |

| 17 | 18 | 19 | 20 | 21 | 22 | 23 | 24 |
|:---:|:---:|:---:|:---:|:---:|:---:|:---:|:---:|
| 386(1) | 386(1) | 402(1) | 402(1) | 402(2) | 402(3) | 417(1) | 417(3) |

| 25 | 26 | 27 |
|:---:|:---:|:---:|
| 433(1) | 495(1) | 557(1) |

Rank sums:

$$R_{1.} = 7.5+7.5+11+17.5+17.5+20.5+20.5+23.5+25+26+27 = 203.5,$$

$$R_{2.} = 2 + 4.5 + 4.5 + 7.5 + 11 + 14 + 16 + 20.5 = 80,$$

$$R_{3.} = 1 + 3 + 7.5 + 11 + 14 + 14 + 20.5 + 23.5 = 94.5.$$

Samples sizes: $n_1 = 11$, $n_2 = 8$, $n_3 = 8$; the size of the unified sample is $n = 27$.

The Kruskal–Wallis statistic's value is

$$F_{KW} = \frac{12}{27 \cdot 28} \left( \frac{203.5^2}{11} + \frac{80^2}{8} + \frac{94.5^2}{8} \right) - 3 \cdot 28 = 6.175$$

There are $s = 7$ coincident ranks groups of sizes

$$t_1 = 2, \, t_2 = 4, \, t_3 = 3, \, t_4 = 3, \, t_5 = 2,$$

$$t_6 = 4, \, t_7 = 2$$

so

$$T_1 = 8 - 2 = 6, \, T_2 = 60, \, T_3 = 24, \, T_4 = 24, \, T_5 = 6,$$

$$T_6 = 64 - 4 = 60, \, T_7 = 6$$

The modified Kruskal–Wallis statistic is

$$F_{KW}^* = \frac{6.175}{1 - (3 \cdot 6 + 2 \cdot 24 + 2 \cdot 60)/(27(729 - 1))} =$$

$$\frac{6.175}{0.99054} = 6.234$$

The $P$-value computed using the SPSS package is $pv = 0.03948$.

The asymptotic $P$-value is

$$pv_a = \mathbf{P}\{\chi_2^2 > 6.234\} \approx 0.0443$$

If the significance level is 0.05 then the zero hypothesis is rejected, by exact and by asymptotic tests.

**Comment 4.16.**   Under the scale alternative:

$$H_1 : F_j(x) = F_j(x/\theta_j), \quad x \in \mathbf{R}, \quad \theta_j > 0, \quad j = 1, ..., k$$

$$\exists \, \theta_i \neq \theta_j, \quad 1 \leq i \neq j \leq k \qquad [4.72]$$

and the Kruskal–Wallis test is not powerful.

As in section 4.4, instead of the ranks $R_{ij}$, take $s(R_{ij})$, where $s(r)$ is a function defined on the set $\{1, 2, ..., n\}$ and the statistic

$$F = \frac{1}{\sigma_s^2} \sum_{i=1}^{k} n_i(\bar{s}_i - \bar{s})^2, \quad \bar{s}_i = \frac{1}{n_i} \sum_{j=1}^{n_i} s(R_{ij})$$

$$\bar{s} = \frac{1}{n} \sum_{i=1}^{k} \sum_{j=1}^{n_i} s(R_{ij}), \quad \sigma_s^2 = \frac{1}{n-1} \sum_{i=1}^{k} \sum_{j=1}^{n_i} (s(R_{ij}) - \bar{s})^2 \quad [4.73]$$

Taking the same functions as in section 4.4 we obtain the analogs of the Siegel–Tukey, Ansari–Bradley, Mood and Klotz statistics $F_{ZT}, F_{AB}, F_M, F_K$ for $k > 2$ samples.

Under the hypothesis $H_0$, the distribution of the statistic $F_{ZT}$ coincides with the distribution of $F_{KW}$.

For small sample sizes $n_i$, the $P$-values of the statistics $F_{ZT}, F_{AB}, F_M, F_K$ are computed by statistical software.

As in Theorem 4.13, it is proved that other statistics $F_{KW}$, $F_{AB}, F_M, F_K$ are asymptotically chi-squared distributed with $k - 1$ degrees of freedom.

For large samples, the hypothesis $H_0$ is rejected against the location alternative with an approximate significance level $\alpha$ if

$$F > \chi_\alpha^2(k - 1) \qquad [4.74]$$

where $F$ is any of the statistics $F_{ZT}, F_{AB}, F_M, F_K$.

**Example 4.12.** (continuation of Example 4.7). Fifteen independent measurements of the same quantity by each of two different devices were given in Example 4.7. Using two other devices, the following results were obtained:

device 3: 2.0742; -1.0499; 0.8555; 3.2287; 2.2756; 1.4305; 0.8750; -0.5559; -1.2684; -0.4667; 1.0099; -2.9228; -0.1835; 1.0803; 0.0759.

device 4: 1.7644; 0.4839; -2.1736; 0.9326; -1.0432; -0.1026; 0.9777; 0.6117; -0.4034; 2.6000; -0.9851; 0.0052; -0.5035; -2.7274; 0.5828.

Unifying this data with the data in Example 4.7 lets us test the hypothesis that the accuracy of all four devices is the same.

If we suppose that independent normal zero mean random variables are observed then the hypothesis is equivalent to the parametric hypothesis $H : \sigma_i^2 = \sigma_2^2 = \sigma_3^2 = \sigma_4^2$ on the equality of variances. This hypothesis may be tested using the Bartlett test, based on the likelihood ratio statistic

$$R_{TS} = n\ln(s^2) - \sum_{i=1}^{4} n_i \ln(s_i^2)$$

where $n = n_1 + n_2 + n_3 + n_4$, $s_i^2$ is the estimator of the variance from the $i$-th samples given in Example 4.6, and $s^2 = (n_1 s_1^2 + ... + n_4 s_4^2)/n$. We have $R_{TS} = 6.7055$ and the asymptotic $P$-value is $pv_a = \mathbf{P}\{\chi_3^2 > 6.7055\} = 0.0819$.

Let us now apply the non-parametric tests given in section 4.5. The values of the statistics are: $S_{ZT} = 8.9309, S_{AB} = 8.8003, S_M = 4.9060, S_K = 6.6202$. Using the SAS statistical package we obtained the asymptotic $P$-values 0.0302, 0.0321, 0.1788 and 0.0850, respectively. Note that the parametric test give an answer near to that obtained using the Klotz test.

## 4.9. Homogeneity hypotheses for $k$ related samples: Friedman test

Suppose that $n$ independent random vectors

$$(X_{11}, ..., X_{1k})^T, ..., (X_{n1}, ..., X_{nk})^T, \quad k > 2$$

are observed. The cdf $F_i = F_i(x_1, ..., x_k)$ of the random vector $\mathbf{X}_i = (X_{i1}, ..., X_{ik})^T$ belongs to a non-parametric class $\mathcal{F}$ of $k$-dimensional absolutely continuous distributions. So $X_{ij}$ is the $j$-th coordinate of the $i$-th random vector.

The data can also be interpreted as $k$ related (dependent) vectors

$$(X_{11}, ..., X_{n1})^T, ..., (X_{1k}, ..., X_{nk})^T$$

where $\mathbf{Y}_j = (X_{1j}, ..., X_{nj})^T$ is a vector with independent coordinates.

Let us write all the observations in the following matrix:

$$\mathbf{X} = \begin{pmatrix} X_{11} & X_{12} & \cdots & X_{1k} \\ X_{21} & X_{22} & \cdots & X_{2k} \\ \cdots & \cdots & \cdots & \cdots \\ X_{n1} & X_{n2} & \cdots & X_{nk} \end{pmatrix}. \qquad [4.75]$$

Denote by $F_{i1}, ..., F_{ik}$ the marginal cumulative distribution functions of the random vector $\mathbf{X}_i = (X_{i1}, ..., X_{ik})^T$.

**Homogeneity hypothesis of $k$ related samples:**

$$H_0 : F_{i1}(x) = ... = F_{ik}(x), \quad \forall x \in \mathbf{R}, \quad i = 1, ..., n$$

Let us consider several situations where such a hypothesis arises.

**Example 4.13.** [BRO 60] Each day of the week ($k = 7$) the work efficiency of $n = 6$ groups of workers was measured. The random variable $X_{ij}$ is the efficiency of the $i$-th group during the $j$-th day of the week. The coordinates of the random vector $\mathbf{X}_i = (X_{i1}, ..., X_{ik})^T$ are dependent (the same group is observed $k$ times) but not necessarily equally distributed. The vectors $\mathbf{X}_1, ..., \mathbf{X}_n$ are independent (different groups of workers). So the coordinates of the vector $\mathbf{Y}_j = (X_{1j}, ..., X_{nj})^T$ are independent (different groups of workers) and may be equally distributed (if groups have equal qualifications) or differently distributed (if groups have different qualifications). The random vectors $Y_1,...,Y_k$ are dependent (observations of the same groups on different days).

The problem is to verify the hypothesis that efficiency does not depend on the day of the week. In this problem, not groups but days are compared.

**Example 4.14.** The reaction time of $n = 10$ patients to $k = 3$ types of medications is fixed. The random variable $X_{ij}$ is the reaction time of the $i$-th patient after application of the $j$-th medication. The problem is to verify the hypothesis that the influence of all medications on the reaction time is the same.

Suppose that $F_i$ belongs to the class $\mathcal{F}_i$, the class of absolutely continuous cdf of the form

$$F_i(x_1, x_2, \ldots, x_k) = G_i(x_1, x_2 + \theta_{i2}, \ldots, x_k + \theta_{ik}), \quad \theta_{ij} \in \mathbf{R}$$

$G_i$ are symmetric functions, i.e. for all $x_1, \ldots, x_k \in \mathbf{R}$, and for any permutation $(j_1, \ldots, j_k)$ of $(1, \ldots, k)$

$$G_i(x_{j_1}, \ldots, x_{j_k}) = G_i(x_1, \ldots, x_k) \qquad [4.76]$$

The alternative $H_1$ to the homogeneity hypothesis $H_0$ is written as follows

$$H_1 : \ F_i \in \mathcal{F}, \quad F_{ij}(x) = F_{i1}(x - \theta_{ij}), \ i = 1, \ldots, n; \ j = 2, \ldots, k;$$

$$\sum_{i=1}^{n} \sum_{j=2}^{k} \theta_{ij}^2 > 0$$

i.e. the marginal cdf differ only in location parameter.

**Construction of the Friedman test statistic.** Denote by $(R_{i1}, ..., R_{ik})^T$ the rank vector of the $i$-th vector $(X_{i1}, ..., X_{ik})^T$.

Replace the data matrix $\mathbf{X}$ with the ranks matrix:

$$
\boldsymbol{R} = \begin{pmatrix}
R_{11} & R_{12} & \cdots & R_{1k} \\
R_{21} & R_{22} & \cdots & R_{2k} \\
\cdots & \cdots & \cdots & \cdots \\
R_{n1} & R_{n2} & \cdots & R_{nk}
\end{pmatrix}
$$

The sum of the ranks in each line is the same:

$$
R_{i\cdot} = R_{i1} + \ldots + R_{ik} = k(k+1)/2, \quad i = 1, 2, \ldots, n.
$$

Find the mean rank in the $j$-th columns:

$$
\bar{R}_{\cdot j} = \frac{1}{n} \sum_{i=1}^{n} R_{ij}
$$

The arithmetic mean of the ranks of all observations is:

$$
\bar{R}_{\cdot\cdot} = \frac{1}{nk} \sum_{i=1}^{n} \sum_{j=1}^{k} R_{ij} = \frac{k+1}{2}
$$

Using the symmetry of distributions, under the hypothesis $H_0$, the random variables $\bar{R}_{\cdot 1}, \ldots, \bar{R}_{\cdot k}$ are identically distributed and are spread around the mean rank $\bar{R}_{\cdot\cdot}$.

The Friedman test statistic is based on the differences $\bar{R}_{\cdot j} - \bar{R}_{\cdot\cdot} = \bar{R}_{\cdot j} - (k+1)/2$

$$
S_F = \frac{12n}{k(k+1)} \sum_{j=1}^{k} \left(\bar{R}_{\cdot j} - \frac{k+1}{2}\right)^2 = \frac{12n}{k(k+1)} \sum_{j=1}^{k} \bar{R}_{\cdot j}^2 - 3n(k+1) =
$$

$$
= \frac{12}{nk(k+1)} \sum_{j=1}^{k} R_{\cdot j}^2 - 3n(k+1) \tag{4.77}
$$

where $R_{\cdot j} = \sum_{i=1}^{n} R_{ij}$. The normalizing factor $12n/k(k+1)$ is chosen to give the asymptotic chi-squared distribution (as $n \to \infty$) of the statistic $S_F$ under the hypothesis $H_0$.

**Friedman test:** the hypothesis $H_0$ is rejected if $S_F \geq S_{F,\alpha}$, where $S_{F,\alpha}$ is the smallest real number $c$ verifying the inequality $\mathbf{P}\{S_F \geq c | H_0\} \leq \alpha$.

The $P$-value is $pv = \mathbf{P}\{S_F \geq s\}$, where $s$ is the observed value of the statistic $S_F$. If $n$ is not large then the critical values of the Friedman statistic are computed using the distribution of ranks described in section 4.1.

**Large samples.**

Let us find the limit distribution of the statistic as $n \to \infty$.

**Theorem 4.14.** *Under the hypothesis* $H_0$

$$S_F \xrightarrow{d} S \sim \chi^2(k-1), \quad n \to \infty \qquad [4.78]$$

**Proof.** Set

$$\bar{\mathbf{R}} = (\bar{R}_{\cdot 1}, \dots, \bar{R}_{\cdot k})^T$$

From [4.3], the mean $\mathbf{E}(\bar{\mathbf{R}})$ and the covariance matrix $\Sigma_n$ of the random vector $\bar{\mathbf{R}}$ are

$$\mathbf{E}(\bar{\mathbf{R}}) = ((k+1)/2, \dots, (k+1)/2), \quad \Sigma_n = [\sigma_{ls}]_{k \times k}$$

where

$$\sigma_{ls} = \mathbf{Cov}(\bar{R}_{\cdot l}, \bar{R}_{\cdot s}) = \frac{1}{n}\mathbf{Cov}(R_{1l}, R_{1s}) =$$

$$\begin{cases} (k^2 - 1)/(12n), & \text{if } l = s, \\ -(k+1)/(12n), & \text{if } l \neq s. \end{cases}$$

So

$$\Sigma_n = \frac{k(k+1)}{12n}\left(\mathbf{E}_k - \frac{1}{k}\mathbf{1}\mathbf{1}^T\right)$$

where $\mathbf{E}_k$ is the unit matrix, $\mathbf{1} = (1, \dots, 1)^T$. The rank of the matrix $\Sigma_n$ is $k-1$ because

$$k\mathbf{E}_k - \mathbf{1}\mathbf{1}^T = \begin{pmatrix} k-1 & -1 & \cdots & -1 \\ -1 & k-1 & \cdots & -1 \\ \cdots & \cdots & \cdots & \cdots \\ -1 & -1 & \cdots & k-1 \end{pmatrix} \sim$$

$$\begin{pmatrix} k-1 & -1 & \cdots & -1 \\ -k & k & \cdots & 0 \\ \cdots & \cdots & \cdots & \cdots \\ -k & 0 & \cdots & k \end{pmatrix} \sim \begin{pmatrix} 0 & -1 & \cdots & -1 \\ 0 & k & \cdots & 0 \\ \cdots & \cdots & \cdots & \cdots \\ 0 & 0 & \cdots & k \end{pmatrix}.$$

We subtracted the first line from the other lines and then added the other columns to the first column.

Using the CLT for sequences of random vectors

$$\sqrt{n}(\bar{\mathbf{R}} - \mathbf{E}(\bar{\mathbf{R}})) \overset{d}{\to} \mathbf{Z} \sim N_k(\mathbf{0}, \mathbf{\Sigma})$$

where

$$\mathbf{\Sigma} = n\mathbf{\Sigma}_n = k(k+1)(\mathbf{E}_k - \frac{1}{k}\mathbf{1}\mathbf{1}^T)/12, \quad Rank(\mathbf{\Sigma}) = k-1$$

As in the case of Kruskal–Wallis's test, use the theorem stating that if $\mathbf{X} \sim N_k(\boldsymbol{\mu}, \mathbf{\Sigma})$ then $(\mathbf{X} - \boldsymbol{\mu})^T\mathbf{\Sigma}^-(\mathbf{X} - \boldsymbol{\mu}) \sim \chi^2(r)$, where $\mathbf{\Sigma}^-$ is the generalized inverse of $\mathbf{\Sigma}$, $r$ is the rank of the matrix $\mathbf{\Sigma}$. By this theorem

$$Q_n = \sqrt{n}(\bar{\mathbf{R}} - \mathbf{E}\bar{\mathbf{R}})^T\mathbf{\Sigma}^-\sqrt{n}(\bar{\mathbf{R}} - \mathbf{E}\bar{\mathbf{R}}) \overset{d}{\to} S \sim \chi^2(k-1).$$

Let us show that $Q_n = S_F$. We obtain

$$(\mathbf{E}_k - \frac{1}{k}\mathbf{1}\mathbf{1}^T)^- = \mathbf{E}_k + k\mathbf{1}\mathbf{1}^T$$

because

$$\mathbf{1}^T\mathbf{1}\mathbf{1}^T = k\mathbf{1}\mathbf{1}^T$$

and

$$(\mathbf{E}_k - \frac{1}{k}\mathbf{1}\mathbf{1}^T)(\mathbf{E}_k + k\mathbf{1}\mathbf{1}^T)(\mathbf{E}_k - \frac{1}{k}\mathbf{1}\mathbf{1}^T) = (\mathbf{E}_k - \frac{1}{k}\mathbf{1}\mathbf{1}^T)(\mathbf{E}_k - \frac{1}{k}\mathbf{1}\mathbf{1}^T) =$$

$$(\mathbf{E}_k - \frac{1}{k}\mathbf{1}\mathbf{1}^T)$$

Hence

$$\mathbf{\Sigma}^- = \frac{12}{k(k+1)}(\mathbf{E}_k + k\mathbf{1}\mathbf{1}^T)$$

and

$$Q_n = \frac{12n}{k(k+1)} \left( \sum_{j=1}^{k} (\bar{R}_{\cdot j} - \frac{k+1}{2})^2 + k[\sum_{j=1}^{k} (\bar{R}_{\cdot j} - \frac{k+1}{2})]^2 \right) =$$

$$\frac{12n}{k(k+1)} \sum_{j=1}^{k} \left( \bar{R}_{\cdot j} - \frac{k+1}{2} \right)^2 = S_F$$

$\triangle$

**Asymptotic Friedman test:** if $n$ is large then the hypothesis $H_0$ is rejected with an approximate significance level $\alpha$ if

$$S_F > \chi_\alpha^2(k-1) \qquad\qquad [4.79]$$

So, the approximate $P$-value is $1 - F_{\chi_{k-1}^2}(s)$, where $s$ is the observed value of the statistic $S_F$.

**Data with** *ex aequo* Let us prove the generalization of Theorem 4.2 to the case where distributions are not necessarily continuous. Set

$$S_F^* = \frac{S_F}{1 - \sum_{i=1}^{n} T_i/(n(k^3 - k))}$$

where

$$T_i = \sum_{j=1}^{k_i} (t_{ij}^3 - t_{ij})$$

$k_i$ is the number of groups with coinciding values for the $i$-th object (i.e. in the $i$-th line of matrix [4.75]), $t_{ij}$ is the size of the $j$-th group.

**Theorem 4.15.** *Under the hypothesis $H_0$*

$$S_F^* \xrightarrow{d} S \sim \chi^2(k-1), \quad n \to \infty \qquad\qquad [4.80]$$

**Proof.** The proof is similar to the proof of Theorem 4.13, so we use the same notation. Set $t = \mathbf{E}T_i$. From [4.6] the means of the components of the random vector $\bar{\mathbf{R}}$ are $(k+1)/2$ and the elements of the covariance matrices $\Sigma_n$ are:

$$\sigma_{ls} = \begin{cases} \frac{1}{12n}(k^2 - 1 - \frac{t}{k}), & \text{if } l = s, \\ \frac{1}{12n}(-(k+1) + \frac{t}{k(k-1)}), & \text{if } l \neq s. \end{cases} =$$

$$\begin{cases} \frac{k(k+1)}{12n}(1 - \frac{t}{k^3-k})(1 - \frac{1}{k}), & \text{if } l = s, \\ \frac{k(k+1)}{12n}(1 - \frac{t}{k^3-k})(-\frac{1}{k}), & \text{if } l \neq s. \end{cases}$$

So

$$\Sigma_n = \frac{k(k+1)}{12n}(1 - \frac{t}{k^3-k})(\mathbf{E}_k - \frac{1}{k}\mathbf{1}\mathbf{1}^T)$$

$$\Sigma^- = \frac{12}{k(k+1)}(1 - \frac{t}{k^3-k})(\mathbf{E}_k + k\mathbf{1}\mathbf{1}^T)$$

and hence

$$\frac{S_F}{1 - \frac{t}{k^3-k}} \xrightarrow{d} S \sim \chi^2(k-1)$$

The convergence of the distribution of the statistic $S_F^*$ to the chi-squared distribution is obtained by replacing $t$ with $\frac{1}{n}\sum_{i=1}^n T_i$ because

$$\frac{1}{n}\sum_{i=1}^n T_i \xrightarrow{P} t = \mathbf{E}T_1$$

$\triangle$

**Example 4.15.** Using $k = 3$ methods, the quantity of amylase is determined for the same $n = 9$ pancreatitis patients. The data are given in the following table.

| Patient | Method 1 | 2 | 3 |
|---|---|---|---|
| 1 | 4,000 | 32,10 | 6,120 |
| 2 | 1,600 | 1,040 | 2,410 |
| 3 | 1,600 | 647 | 2,210 |
| 4 | 1,200 | 570 | 2,060 |
| 5 | 840 | 445 | 1,400 |
| 6 | 352 | 156 | 249 |
| 7 | 224 | 155 | 224 |
| 8 | 200 | 99 | 208 |
| 9 | 184 | 70 | 227 |

Verify the hypothesis that all three methods are equivalent.

The quantities of amylase determined for different patients using a specified method are independent random variables. They may be differently distributed (for example if the durations of the disease are different). For a specific patient, the quantities of amylase determined by different methods are dependent random variables.

The table of ranks is:

| Patient | Method 1 | 2 | 3 |
|---|---|---|---|
| 1 | 2 | 1 | 3 |
| 2 | 2 | 1 | 3 |
| 3 | 2 | 1 | 3 |
| 4 | 2 | 1 | 3 |
| 5 | 2 | 1 | 3 |
| 6 | 3 | 1 | 2 |
| 7 | 2.5 | 1 | 2.5 |
| 8 | 2 | 1 | 3 |
| 9 | 2 | 1 | 3 |
| | $R_{.1} = 19.5$ | $R_{.2} = 9$ | $R_{.3} = 25.5$ |

Since $n = 9$, $k = 3$

$$S_F = \frac{12}{9 \cdot 3 \cdot 4}(19.5^2 + 9^2 + 25.5^2) - 3 \cdot 9 \cdot 4 = 15.5$$

In the case of the seventh patient there is one group, of size two, of *ex aequo*

$$k_7 = 1, \ t_7 = 2, \ T_7 = 2^3 - 2 = 6, \ \sum_{i=1}^{9} T_i = 6$$

so

$$1 - \frac{\sum_{i=1}^{n} T_i}{nk(k^2 - 1)} = 1 - \frac{6}{9 \cdot 3 \cdot 8} = 0.97222$$

and

$$S_F^* = \frac{S_F}{0.97222} = 15.943$$

The exact $P$-value is $pv = 0.00034518$. The hypothesis is rejected with a very small significance level. The asymptotic $P$-value, $pv_a = 1 - F_{\chi_2^2}(15.943) = 0.00001072$, differs considerably from the exact value because the samples are not large.

**Example 4.16.** (continuation of Example 4.13). Each day of the week ($k = 7$) the work efficiency of $n = 6$ groups of workers was measured. The data are given in the following table.

| Group | Day | | | | | | |
|-------|-----|-----|-----|-----|-----|-----|-----|
|       | 1   | 2   | 3   | 4   | 5   | 6   | 7   |
| 1     | 60  | 62  | 58  | 52  | 31  | 23  | 26  |
| 2     | 64  | 63  | 63  | 36  | 34  | 32  | 27  |
| 3     | 14  | 46  | 47  | 39  | 42  | 43  | 57  |
| 4     | 30  | 41  | 40  | 41  | 37  | 17  | 12  |
| 5     | 72  | 38  | 46  | 47  | 38  | 60  | 41  |
| 6     | 35  | 35  | 33  | 46  | 47  | 47  | 38  |

Test the hypothesis that the work efficiency does not depend on the day of the week.

The table of ranks is as follows:

| Group | Day | | | | | | |
|---|---|---|---|---|---|---|---|
| | 1 | 2 | 3 | 4 | 5 | 6 | 7 |
| 1 | 6 | 7 | 5 | 4 | 3 | 1 | 2 |
| 2 | 7 | 5.5 | 5.5 | 4 | 3 | 2 | 1 |
| 3 | 1 | 5 | 6 | 2 | 3 | 4 | 7 |
| 4 | 3 | 6.5 | 5 | 6.5 | 4 | 2 | 1 |
| 5 | 7 | 1.5 | 4 | 5 | 1.5 | 6 | 3 |
| 6 | 2.5 | 2.5 | 1 | 5 | 6.5 | 6.5 | 4 |
| | 26.5 | 28 | 26.5 | 26.5 | 21 | 21.5 | 18 |

Since
$$n = 6, \quad k = 7, \quad S_F =$$
$$(26.5^2 + 28^2 + 26.5^2 + 26.5^2 + 21^2 + 21.5^2 + 18^2)/28 - 3 \cdot 6 \cdot 8 = 3.07143$$

Several groups have coinciding values of work efficiency
$$k_2 = k_4 = k_5 = 1, \quad k_6 = 2, \quad t_2 = t_4 = t_5 = t_{61} = t_{61} = 2$$

$$T_3 = T_4 = T_5 = 6, \quad T_6 = 6 + 6 = 12, \quad \sum_{i=1}^{6} T_i = 30$$

The correction factor is

$$1 - \frac{\sum_{i=1}^{n} T_i}{nk(k^2 - 1)} = 1 - \frac{30}{6 \cdot 7 \cdot 48} = 0.985119$$

so
$$S_F^* = 3.07143/0.985119 = 3.1178$$

The SPSS package does not compute the exact $P$-value because the computation volume is too large. The asymptotic $P$-value is $pv_a = 0.7939$. The data do not contradict the zero hypothesis.

**Comment 4.17.** The Friedman test is also useful in the case where the random vectors $\mathbf{X}_i = (X_{i1}, ..., X_{ik})^T$ are differently distributed but their coordinates are independent. In such a case, the homogeneity hypothesis $H_0$ cannot be verified using the Kruskal–Wallis test because it is based on data with identically distributed $\mathbf{X}_i$, $i = 1, ..., n$.

**Example 4.17.** In [SHE 99], The corrosion resistance of nine alloys was investigated. The alloys were tested under eight different conditions. The data are given in the following table.

|   | 1 | 2 | 3 | 4 | 5 | 6 | 7 | 8 | 9 |
|---|---|---|---|---|---|---|---|---|---|
| 1 | 1.40 | 1.45 | 1.91 | 1.89 | 1.77 | 1.66 | 1.92 | 1.84 | 1.54 |
| 2 | 1.35 | 1.57 | 1.48 | 1.48 | 1.73 | 1.54 | 1.93 | 1.79 | 1.43 |
| 3 | 1.62 | 1.82 | 1.89 | 1.39 | 1.54 | 1.68 | 2.13 | 2.04 | 1.70 |
| 4 | 1.31 | 1.24 | 1.51 | 1.67 | 1.23 | 1.40 | 1.23 | 1.58 | 1.64 |
| 5 | 1.63 | 1.18 | 1.58 | 1.37 | 1.40 | 1.45 | 1.51 | 1.63 | 1.07 |
| 6 | 1.41 | 1.52 | 1.65 | 1.11 | 1.53 | 1.63 | 1.44 | 1.28 | 1.38 |
| 7 | 1.93 | 1.43 | 1.38 | 1.72 | 1.32 | 1.63 | 1.33 | 1.69 | 1.70 |
| 8 | 1.40 | 1.86 | 1.36 | 1.37 | 1.34 | 1.36 | 1.38 | 1.80 | 1.84 |

In this example the coordinates of the vector $(X_1, ..., X_9)^T$ are independent, but, under different conditions, may have different probability distributions. So the hypothesis has the form

$$H_0 : F_{i1}(x) = ... = F_{i9}(x), \quad \forall x \in \mathbf{R}, \quad \forall i = 1, ..., 8$$

The Kruskal–Wallis test cannot be applied here and we use the Friedman test.

We have $S_F = 31.1417$, and. taking into account five *ex aequo*, we have $S_F^* = 31.2720$. The asymptotic $P$-value $pv_a = \mathbf{P}\{\chi_9^2 > 31.2720\} = 0.00027$. The hypothesis is rejected.

**ARE of Friedman's test with respect to Fisher's test: independent samples.** If for any $i$ the random variables $X_{i1}, ..., X_{ik}$ are independent and the alternative is

$$F_{ij}(x) = F(x - \alpha_i - \beta_j), \quad \exists j, j' : \beta_j \neq \beta_{j'}$$

then the hypothesis $H_0$ is equivalent to the equality

$$\beta_1 = \cdots = \beta_k$$

In the parametric case, when the distribution is normal, we get the two-factor ANOVA model, so this hypothesis is verified using Fisher's test with the test statistic

$$F = \frac{n(n-1)\sum_{j=1}^{k}(\bar{X}_{\cdot j} - \bar{X}_{\cdot\cdot})^2}{\sum_{i=1}^{n}\sum_{j=1}^{k}(X_{ij} - \bar{X}_{i\cdot} - \bar{X}_{\cdot j} + \bar{X}_{\cdot\cdot})^2}$$

having the Fisher distribution with $k-1$ and $(n-1)(k-1)$ degrees of freedom under the zero hypothesis. The hypothesis is rejected with a significance level $\alpha$ if

$$F > F_\alpha(k-1, (n-1)(k-1))$$

In the non-parametric case the distribution of the statistic $(k-1)F$ converges to the chi-squared distribution with $k-1$ degrees of freedom as $n \to \infty$, so for large $n$ the hypothesis is rejected with an asymptotic significance level $\alpha$ if

$$(k-1)F > \chi_\alpha^2(k-1)$$

The ARE of Friedman test with respect to Fisher's test is

$$e(S_F, F) = \frac{12k\sigma^2}{k+1}\left[\int_{-\infty}^{\infty} f^2(x)dx\right]^2, \quad f = F', \quad \sigma^2 = \mathbf{Var}(X_{ij})$$

So in the normal case

$$e(S_F, F) = \frac{3k}{(k+1)\pi}$$

The efficiency increases when $k$ increases. If $k = 2$ then the efficiency is rather small, $ARE = 2/\pi \approx 0.637$, i.e. it coincides with the ARE of the sign test considered below. If $k = 5$ then the $ARE = 0.796$ and if $k$ increases then the ARE also increases and approaches 0.955. For distributions with heavy "tails", the ARE of the Friedman test with respect to that of Fisher's test can be greater than 1. We saw that in the case of a doubly exponential distribution the ARE of the Wilcoxon signed ranks test with respect to Student's test is 2, so the ARE of the Friedman test with respect to Fisher's test is $2 - 2/(k + 1)$.

## 4.10. Independence test based on Kendall's concordance coefficient

First let us generalize the Spearman's rank correlation coefficient to the case of $k > 2$ samples.

Suppose that the iid random vectors

$$(X_{11}, ..., X_{1k})^T, ..., (X_{n1}, ..., X_{nk})^T$$

are observed.

Such data are obtained when, for example, $k$ experts rate the quality of $n$ objects: the random vector $(X_{i1}, ..., X_{ik})^T$ contains ratings of the $i$-th object by all experts, $i = 1, ..., n$. The random vector $(X_{1j}, ..., X_{nj})^T$ contains ratings of all objects by the $j$-th expert.

Denote by $(R_{1j}, ..., R_{nj})^T$ the rank vector obtained from the vector $(X_{ij}, ..., X_{nj})^T$. The sum of ranks given by expert $j$ is $R_{.j} = \sum_{i=1}^{n} R_{ij} = n(n + 1)/2$.

The sum of ranks given to the $i$-th object is $R_{i.} = \sum_{j=1}^{k} R_{ij}$. The arithmetic mean of $R_{1.}, ..., R_{n.}$ is

$$\bar{R}_{..} = \frac{1}{n} \sum_{i=1}^{n} R_{i.} = \frac{1}{n} \sum_{i=1}^{n} \sum_{j=1}^{k} R_{ij} = \frac{1}{n} \sum_{j=1}^{k} \sum_{i=1}^{n} R_{ij} =$$

$$\frac{1}{n} k \frac{n(n + 1)}{2} = \frac{k(n + 1)}{2}$$

**Definition 4.7.** The random variable

$$W = \frac{12}{k^2 n(n^2 - 1)} \sum_{i=1}^{n} (R_{i.} - k(n + 1)/2)^2 \qquad [4.81]$$

is called *Kendall's concordance coefficient.*

Denote by

$$R_S^{(jl)} = \frac{12}{n^3 - n} \sum_{i=1}^{n} (R_{ij} - (n + 1)/2)(R_{il} - (n + 1)/2)$$

the Spearman's correlation coefficient of the $j$-th and the $l$-th random vectors $(X_{1j}, ..., X_{nj})^T$ and $(X_{1l}, ..., X_{nl})^T$, i.e. of the ratings of the $j$-th and the $l$-th experts.

**Theorem 4.16.** *Kendall's concordance coefficient is a linear function of the mean of Spearman's correlation coefficients*

$$R_S^{mean} = \frac{2}{k(k - 1)} \sum_{j<l} R_S^{(jl)}$$

*namely*

$$W = \frac{1}{k}(1 + (k - 1)R_S^{mean})$$

**Proof.** We obtain

$$\sum_{i=1}^{n} (R_{i.} - \frac{k(n + 1)}{2})^2 = \sum_{i=1}^{n} \sum_{j=1}^{k} \{(R_{ij} - \frac{n + 1}{2})^2 +$$

$$2 \sum_{j<l} (R_{ij} - \frac{n + 1}{2})(R_{il} - \frac{n + 1}{2})\} =$$

$$= \frac{k(n^3 - n)}{12} + 2\frac{(n^3 - n)}{12} \sum_{j<l} R_S^{(jl)}$$

hence the proposition of the theorem.

The theorem implies that in the case of $k = 2$, the coefficient $W$ is the following linear function of Spearman's correlation coefficient:

$$W = (R_S + 1)/2$$

If all Spearman's correlation coefficients are equal to 1, i.e. the ratings of all experts coincide, then $W$ takes the greatest possible value: $W = 1$.

If all Spearman's correlation coefficients are equal to zero then $W = 1/k$. If they are non-negative then $W \geq 1/k$.

If negative Spearman's correlation coefficients exist then the definition of $W$ and the theorem imply that $W$ can take non-negative values, which can be smaller than $1/k$.

Note that the statistic $W$ can take the value $1/k$ not only in the case when all Spearman's correlation coefficients are equal to zero. For example, if $k = 4$, $X_i = -iX_1$, $i = 2, 3, 4$ then

$$R_S^{(12)} = R_S^{(13)} = R_S^{(14)} = -1, \quad R_S^{(23)} = R_S^{(24)} = R_S^{(34)} = 1$$

so $R_S^{mean} = 0$ and $W = 1/k$.

The concordance coefficient may be used to test the hypothesis of random vector components independence when the alternative is positive correlations of all pairs.

Suppose that the distribution of the random vectors $(X_{i1}, ..., X_{ik})^T$ is absolutely continuous and the correlation coefficients $\rho_{jl} = \mathbf{E}R_S^{(jl)}$ are non-negative.

Denote by $F_i$ the cumulative distribution function of the random vector $X_{i1}, \ldots, X_{ik}$ and by $F_{i1}, ..., F_{ik}$ the marginal cdf.

**Independence hypothesis of random variables** $X_{i1}, \ldots, X_{ik}$ **:**

$$H_0 : F_i(x_1, \ldots, x_k) = F_{11}(x_1) \ldots F_{ik}(x_k),$$

$$\forall x_1, \ldots, x_k \in \mathbf{R}, \quad i = 1, \ldots, n$$

Suppose that the alternative is

$$\bar{H} : \rho_{jl} \geq 0, \ \forall \, 1 \leq j < l \leq k \,, \ \exists \rho_{i_0, l_0} > 0$$

**Independence test based on the concordance coefficient:** the hypothesis $H_0$ is rejected if $W > W_\alpha$, where $W_\alpha$ is the minimum real number $c$ verifying the inequality $\mathbf{P}\{W > c | H_0\} \leq \alpha$.

Note that the statistic $k(n - 1)W$ is computed using the same formula as for the Friedman statistic if the roles of $n$ and $k$ are exchanged. So, under the independence hypothesis, the distribution of $k(n - 1)W$ coincides with the distribution of the Friedman statistic when the roles of $n$ and $k$ are exchanged. So, the critical values of the Friedman statistic can be used. The distribution of the statistic $W$ is asymmetric in the interval $[0, \ 1]$. Using the relation to Friedman statistic, we see that if $k$ is large then, under the independence hypothesis, the distribution of $k(n-1)W$ is approximated by a chi-squared distribution with $n - 1$ degrees of freedom.

If $k$ is not large then other approximations are used. For example, using the fact that under the independence hypothesis

$$\mathbf{E}W = \frac{1}{k}, \quad \mathbf{Var}(W) = \frac{2(k - 1)}{k^3(n - 1)}$$

and the distribution of the statistics $W$ is concentrated in the interval $[0, \ 1]$, this distribution is approximated by the beta distribution with the same first two moments

$$W \approx Y \sim Be(\gamma, \eta), \quad \gamma = \frac{n - 1}{2} - \frac{1}{k}, \quad \eta = (k - 1)\gamma \quad [4.82]$$

**Example 4.18.** The notes of 20 students in probability, mathematical analysis and statistics $X_{1i}, X_{2i}, X_{3i}, i = 1, ..., 20$, are given in the following table.

| $i$ | 1 | 2 | 3 | 4 | 5 | 6 | 7 | 8 | 9 | 10 | 11 | 12 | 13 | 14 | 15 | 16 | 17 | 18 | 19 | 20 |
|---|---|---|---|---|---|---|---|---|---|---|---|---|---|---|---|---|---|---|---|---|
| $X_{1i}$ | 7 | 5 | 6 | 6 | 8 | 4 | 5 | 5 | 8 | 8 | 7 | 5 | 6 | 3 | 7 | 7 | 8 | 7 | 6 | 5 |
| $X_{2i}$ | 7 | 5 | 8 | 5 | 8 | 3 | 5 | 4 | 7 | 6 | 6 | 3 | 7 | 4 | 5 | 8 | 7 | 5 | 5 | 5 |
| $X_{3i}$ | 9 | 5 | 6 | 5 | 8 | 5 | 5 | 5 | 8 | 7 | 8 | 5 | 5 | 5 | 7 | 7 | 7 | 6 | 7 | 5 |

Test the hypothesis that the notes are independent against the alternative that $X_1, X_2, X_3$ are positively correlated. Using the SPSS package, we have $W = 0.859$ and $pv_a = 0.00026$. The hypothesis is rejected.

## 4.11. Bibliographic notes

Rank tests for independence were first proposed by Spearman [SPE 04] and Kendall [KEN 38, KEN 39, KEN 62]. Rank tests for homogeneity (independent samples): Wilcoxon [WIL 45, WIL 47], Mann and Whitney [MAN 47], Van der Vaerden [VAN 52], Kruskal and Wallis [KRU 52] ($k > 2$ samples), Mood [MOO 54], Ansari and Bradley [ANS 60], Siegel and Tukey [SIE 60] and Klotz [KLO 65].

Rank tests for homogeneity (related samples) were first proposed by Wilcoxon [WIL 45, WIL 47] (two samples) and Friedman [FRI 37] ($k > 2$ samples).

Bartels–Von Neuman's rank test for randomness was proposed in [BAR 82].

The asymptotic theory for rank tests is discussed in Hájek and Sidák [HAJ 67] and Hájek *et al.* [HAJ 99].

A fuller survey of results on rank tests can be found in Hettmansperger [HET 84], Hollander and Wolfe [HOL 99], Sprent and Smeeton [SPR 01], Gibbons and Chakraborti [GIB 09] and Govindarajulu [GOV 07].

## 4.12. Exercises

**4.1.** Prove that in the case of an absolutely continuous distribution the Spearman correlation coefficient $r_S = 1 - 12V/(n(n^2 - 1))$, where $V = \sum_{i<j} h_{ij}(j - i)$, $h_{ij} = 1$, if $R_i > R_j$ and $h_{ij} = 0$, if $R_i < R_j$.

**4.2.** Prove that the Pearson correlation coefficient of the statistics $R_S$ and $R_K$ is

$$\rho(R_S, R_K) = 2(n + 1)/\sqrt{2n(2n + 5)}$$

**4.3.** The quantity of starch was determined in 16 potatoes using two methods. The results are given in the table ($X_i$ is the first method, $Y_i$ is the second method).

| $i$ | $X_i$ | $Y_i$ | $i$ | $X_i$ | $Y_i$ |
|---|---|---|---|---|---|
| 1 | 21.7 | 21,5 | 9 | 14.0 | 13,9 |
| 2 | 18.7 | 18.7 | 10 | 17,2 | 17,0 |
| 3 | 18.3 | 18.3 | 11 | 21.7 | 21,4 |
| 4 | 17,5 | 17.4 | 12 | 18,6 | 18,6 |
| 5 | 18.5 | 18.3 | 13 | 17.9 | 18,0 |
| 6 | 15.6 | 15.4 | 14 | 17.7 | 17.6 |
| 7 | 17,0 | 16.7 | 15 | 18.3 | 18.5 |
| 8 | 16.6 | 16,9 | 16 | 15.6 | 15,5 |

Test the independence hypothesis of the random variables $X$ and $Y$ using the Spearman and Kendall rank correlation coefficients.

**4.4.** In investigating the efficiency of a new seeder, ten lots were seeded using the usual seeder and ten lots were seeded using the new seeder. The fertility of these lots was compared. To eliminate the influence of the soil, the 20 lots of equal area were adjacent. One of the lots of the pair was seeded using the usual method, and the other was seeded by the new method. The results are given in the table ($X_i$ is the fertility of the $i$-th lot seeded using the new seeder, $Y_i$ is the fertility using the usual seeder).

| $i$ | $X_i$ | $Y_i$ | $i$ | $X_i$ | $Y_i$ |
|----|-----|-----|----|-----|-----|
| 1 | 8,0 | 5.6 | 6 | 7.7 | 6,1 |
| 2 | 8,4 | 7.4 | 7 | 7.7 | 6.6 |
| 3 | 8,0 | 7.3 | 8 | 5.6 | 6.0 |
| 4 | 6,4 | 6,4 | 9 | 5.6 | 5,5 |
| 5 | 8,6 | 7,5 | 10 | 6.2 | 5,5 |

Test the independence hypothesis of the random variables $X$ and $Y$ using the Spearman and Kendall rank correlation coefficients.

**4.5.** Test the randomness hypothesis using the data in exercise 2.16.

**4.6.** Test the randomness hypothesis using the data in exercise 2.17.

**4.7.** Apply the Wilcoxon and Van der Varden tests to test the hypothesis of the equal effects of poisons, using the data in exercise 3.14.

**4.8.** Prove that in the case of the gama distribution $G(1, \eta)$ the ARE of the Wilcoxon rank sum test with respect to the Student test under location alternatives is

$$A_{W,t} = A(\eta) = \frac{3\eta}{2^{4(\eta-1)}(2(\eta-1)B(\eta,\eta))^2}$$

Show that $A(\eta) > 1,25$ if $\eta \leq 3$; $A(\eta) \to \infty$ if $\eta \to 1/2$; $A(\eta) \to 3/\pi$ if $\eta \to \infty$.

**4.9.** Two experiments were performed. An impulse of intensity 10 was evaluated as 9, 9, 8, 10, 12. 13, 10, 11 units; an impulse of intensity 20 as 15, 16, 17, 23, 22. 20, 21, 24, 27. Apply the rank tests of section 4.5.5 to test the hypothesis that the impulse recognition error does not depend on the impulse intensity against scale alternatives.

**4.10.** Divide the data in exercise 2.16 into ten samples of equal size (the data are given in different lines). Apply the Kruskal–Wallis test and the test based on inversions to test the homogeneity hypothesis for these ten samples.

**4.11.** [CHE 04] TV tubes produced in three plants where tested. Their first failure times (in months) are given in the table. Is the distribution of the failure times of the TV tubes the same in all three plants?

| | |
|---|---|
| Plant 1 tubes | 41 70 26 89 62 54 46 77 34 51 |
| Plant 2 tubes | 30 69 42 60 44 74 32 47 45 37 52 81 |
| Plant 3 tubes | 23 35 29 38 21 53 31 25 36 50 61 |

**4.12.** [CHE 04] Five independent experts rate beer of three sorts. The data are given in the following table.

| Expert/beer | A | B | C |
|---|---|---|---|
| First | 10 | 7 | 8 |
| Second | 5 | 2 | 4 |
| Third | 6 | 8 | 6 |
| Fourth | 3 | 4 | 6 |
| Fifth | 9 | 8 | 10 |

Are the ratings of experts independent?

**4.13.** Apply the Wilcoxon signed ranks test for two dependent samples to test the hypothesis of the equality of distributions of the random variables $X$ and $Y$, using the data in exercise 4.3.

**4.14.** Apply the Wilcoxon signed ranks test for two dependent samples to test the hypothesis of the equality of distributions of the random variables $X$ and $Y$, using the data in exercise 4.4.

**4.15.** [CHE 04] In the table, data on the prices of 12 printers of different types, proposed by three suppliers, are given.

| Type | 1 | 2 | 3 | Type | 1 | 2 | 3 |
|------|------|------|------|------|------|------|------|
| 1 | 660 | 673 | 658 | 7 | 1980 | 1950 | 1970 |
| 2 | 790 | 799 | 785 | 8 | 2300 | 2295 | 2310 |
| 3 | 590 | 580 | 599 | 9 | 2500 | 2480 | 2490 |
| 4 | 950 | 945 | 960 | 10 | 2190 | 2190 | 2210 |
| 5 | 1290 | 1280 | 1295 | 11 | 5590 | 5500 | 5550 |
| 6 | 1550 | 1500 | 1499 | 12 | 6000 | 6100 | 6090 |

Apply the Friedman test to test the hypothesis that the different suppliers' printer prices are equally distributed.

## 4.13. Answers

**4.3.** The values of the Spearman and Kendall correlation coefficients are $r_S = 0.9889$ and $\tau_b = 0.9492$; for both tests $P$-values are $pv < 0.0001$. The independence hypothesis is rejected.

**4.4.** The values of the Spearman and Kendall correlation coefficients are $r_S = 0.7822$ and $\tau_b = 0.6746$; the $P$-values are

$pv = 0.0075$ and $pv = 0.0084$, respectively. The independence hypothesis is rejected.

**4.5.** The values of the Spearman and Kendall correlation coefficients are $r_S = 0.0935$ and $\tau_b = 0.0621$; the $P$-values are $pv = 0.3547$ and $pv = 0.3663$, respectively. The data do not contradict the independence hypothesis.

**4.6.** The values of the Spearman and Kendall correlation coefficients are $r_S = 0.0643$ and $\tau_b = 0.0390$; the $P$-values are $pv = 0.4343$ and $pv = 0.4852$, respectively. The data do not contradict the independence hypothesis.

**4.7.** The value of the Wilcoxon statistic is $W = 239$, $Z_{m,n} = -0.0198$; the asymptotic $P$-value is $pv_a = 2\Phi(-0.0198) = 0.9842$. The value of the Van der Warden statistic is $V = -0.1228$ and $pv_a = 0.9617$. The data do not contradict the homogeneity hypothesis.

**4.9.** The values of the statistics are $S_{ZT} = 91.1667, S_{AB} = 47.5, S_K = 2.2603, S_M = 8.9333$ and the $P$-values are $0.0623$, $0.0713$, $0.0320$ and $0.0342$, respectively. The asymptotic $P$-values are $0.0673, 0.0669, 0.0371$ and $0.0382$, respectively. The homogeneity hypothesis is rejected.

**4.10.** The value of the Kruskal–Wallis statistic is $F_{KW} = 3.9139$ and $pv_a = 0.9170$. The data do not contradict the homogeneity hypothesis.

**4.11.** The value of the Kruskal–Wallis statistic is $F_{KW} = 6.5490$ and $pv_a = 0.0378$. The homogeneity hypothesis is rejected if the significance level is greater than $0.0378$.

**4.12.** The value of the Kendall concordance coefficient is $0.137$ and $pv_a = 0.582, pv_a = 0.504$. The data do not contradict the independence hypothesis.

**4.13.** The value of the signed rank statistic is $T^+ = 69.5$, the $P$-value is $pv = 0.0989$. The homogeneity hypothesis is rejected if the significance level is greater than 0.0989.

**4.14.** The value of the signed rank statistic is $T^+ = 43$ and the $P$-value is $pv = 0.0117$. The homogeneity hypothesis is rejected if the significance level is greater than 0.0117.

**4.15.** The value of the Friedman statistic is $S_F = 2.5957$; the asymptotic $P$-value is $pv_a = 0.2731$. The data do not contradict the hypothesis.

# Chapter 5

# Other Non-parametric Tests

## 5.1. Sign test

### 5.1.1. *Introduction: parametric sign test*

Let us consider $n$ Bernoulli experiments. During an experiment, the event $\{+\}$ or the opposite event $\{-\}$ can occur. Suppose that $n$ experiments are done. Set $p = \mathbf{P}\{+\}$ and denote by $S_1$ and $S_2 = n - S_1$ the number of events $\{+\}$ and $\{-\}$, respectively.

The random variable $S_1$ has the binomial distribution $S_1 \sim B(n, p)$.

The sign test is used to test hypotheses on the equality of the probabilities of the events $\{+\}$ and $\{-\}$, and is based on the number of "successes" $S_1$.

**Hypothesis on the equality of "success" and "failure":**
$H_0 : p = 0.5$.

**Sign test:** under the two-sided alternative $H_3 : p \neq 0.5$, the hypothesis $H_0$ is rejected with a significance level not greater than $\alpha$ if

$$S_1 \leq c_1 \quad \text{or} \quad S_1 \geq c_2 \qquad\qquad [5.1]$$

where $c_1$ is the maximum integer satisfying the equality

$$\mathbf{P}\{S_1 \leq c_1\} = \sum_{k=0}^{c_1} C_n^k (1/2)^n = 1 - I_{0.5}(c_1 + 1, n - c_1) =$$

$$I_{0.5}(n - c_1, c_1 + 1) \leq \alpha/2$$

and $c_2$ is the minimum integer satisfying the equality

$$\mathbf{P}\{S_1 \geq c_2\} = \sum_{k=c_2}^{n} C_n^k (1/2)^n = I_{0.5}(c_2, n - c_2 + 1) \leq \alpha/2$$

where

$$I_{0.5}(a, b) = \frac{1}{B(a, b)} \int_0^{0.5} x^{a-1} (1 - x)^{b-1} dx$$

$$B(a, b) = \int_0^1 x^{a-1} (1 - x)^{b-1} dx$$

If the alternative is one-sided, i.e. $H_1 : p > 0.5$ or $H_2 : p < 0.5$, then the hypothesis is rejected, if $S_1 \geq d_2$ or $S_1 \leq d_1$, respectively; here $d_1$ and $d_2$ verify the same inequalities as $c_1$ and $c_2$, replacing $\alpha/2$ by $\alpha$.

The $P$-value is

$$pv = 2 \min\{I_{0.5}(n - S_1, S_1 + 1), I_{0.5}(S_1, n - S_1 + 1)\}$$

(two-sided alternative $H_3$)

$$pv = I_{0.5}(S_1, n - S_1 + 1) \quad \text{(one-sided alternative } H_1)$$

$$pv = I_{0.5}(n - S_1, S_1 + 1) \quad \text{(one-sided alternative } H_2)$$

**Large sample.** If $n$ is large then the Moivre–Laplace theorem implies

$$(S_1 - n/2)/\sqrt{n/4} \xrightarrow{d} U \sim N(0, 1)$$

Since $S_1 - S_2 = 2S_1 - n$, $S_1 + S_2 = n$

$$Z = \frac{(S_1 - S_2)}{\sqrt{S_1 + S_2}} = \frac{(S_1 - n/2)}{\sqrt{n/4}} \xrightarrow{d} N(0, 1)$$

**Asymptotic sign test:** if $n$ is large then under the two-sided alternative $H_3$, the hypothesis $H_0$ is rejected with an approximate significance level $\alpha$ if $|Z| > z_{\alpha/2}$.

In the case of one-sided alternatives, the critical regions are defined by the inequalities $Z > z_\alpha$ and $Z < -z_\alpha$, respectively.

For medium-sized samples a continuity correction may be used.

**Asymptotic sign test with continuity correction:** under the two-sided alternative, the hypothesis $H_0$ is rejected with an approximate significance level $\alpha$ if

$$Z^* = \frac{(|S_1 - S_2| - 1)^2}{S_1 + S_2} > \chi_\alpha^2(1)$$

**Example 5.1.** [GIB 09] A psychologist interviews the parents of 39 children with learning difficulties to ascertain how well they understand the problems awaiting their children in the future. He established that in $S_1 = 22$ cases the fathers understood the problem better, and in $S_2 = 17$ cases it was the mothers. Are the fathers more understanding than the mothers? The significance level is $\alpha = 0.1$. Solve the same problem if $S_1 = 32$ and $S_2 = 7$.

Test the hypothesis that the mothers are not more understanding than the fathers against the one-sided alternative that the fathers are more understanding. If the hypothesis is rejected we shall be able to say that the data do not contradict the "superiority" of the fathers. Denote by $p$ the probability that the mothers are more understanding than the fathers. The zero hypothesis is $H_0 : p \geq 0.5$ and the alternative is $\bar{H}_2 = p < 0.5$. The hypothesis is rejected if $S_1$ takes small values.

In the first case $n = 39$, $S_1 = 17$ so the $P$-value is

$$pv = I_{0.5}(39 - 17, 17 + 1) = I_{0.5}(22, 18) = 0.261 > 0.1$$

The data do not contradict the hypothesis, so the alternative of the "superiority" of the fathers is rejected. Since

$$Z = (17 - 22)/\sqrt{17 + 22} = -0.8006 > z_{0.9} = -1.282$$

the asymptotic $P$-value is $pv_a = \Phi(-0.8006) = 0.217$. We draw the same conclusion.

In the second case $n = 39$, $S_1 = 7$ so the $P$-value is

$$pv = I_{0.5}(39 - 7, 7 + 1) = I_{0.5}(32, 8) = 0.000035 < 0.1$$

The test rejects the hypothesis, so the data do not contradict the "superiority" of the fathers. Since

$$Z = (7 - 32)/\sqrt{7 + 32} = -4.0032 < z_{0.9} = -1.282$$

the asymptotic $P$-value is $pv_a = \Phi(-4.0032) = 0.000031$.

Let us consider several typical examples of non-parametric hypotheses where the sign test is used.

### 5.1.2. Hypothesis on the nullity of the medians of the differences of random vector components

Suppose that we observe $n$ independent absolutely continuous random vectors

$$(X_1, \ Y_1)^T, \ldots, (X_n, \ Y_n)^T \qquad [5.2]$$

Denote by $\theta_i$ the median of random variable $D_i = X_i - Y_i$.

**Hypothesis on the nullity of the medians of the differences $D_i = X_i - Y_i$:**

$$H_0 : \ \theta_1 = \cdots = \theta_n = 0.$$

This hypothesis is not equivalent to the hypothesis on the equality of the medians of random variables $X_i$ and $Y_i$. There are random vectors with equal medians of the components $X_i$ and $Y_i$ but with non-zero medians of the differences $D_i = X_i - Y_i$.

Under this hypothesis, the distribution of the number of positive differences $S = \sum\limits_{D_i>0} 1$ coincides with the distribution of the parametric sign test statistic $S_1$ considered above. So the exact and the asymptotic tests are defined as in section 5.1.1, replacing $S_1$ by $S$.

If exact values of the differences $D_i = X_i - Y_i$ are known, and the alternative is symmetric with respect to the median of the random variable $D_i$, then usually the Wilcoxon signed ranks test is more powerful because the sum

$$T^+ = \sum_{i:D_i>0} R_i$$

uses not the only signs of the differences $D_i$ but also the values $|D_i|$.

Unlike the Wilcoxon signed ranks test, the sign test can be used in situations where exact values of $D_i$ are not known and *only their signs are known*. For example, we can feel that the pain after usage of some remedy is lesser or greater than the pain after usage of some other remedy, but the pain intensity is not measured. Moreover, the sign test can be used against wider classes of alternatives, supposing that the $D_i$ distribution is not necessarily symmetric.

**Example 5.2.** (continuation of example 4.1). Using the data in Example 4.1, test the hypothesis that the median of the difference of the coordinates of the random vector $(X, Y)^T$ is equal to zero. We know that 50 differences are positive and $S_1 = 28$. Under the two-sided alternative, the $P$-value is

$$pv = 2\min(I_{0.5}(22, 29), \quad I_{0.5}(28, 23)) = 0.4798$$

Using the normal approximation with continuity correction we have $Z^* = (|S_1 - S_2| - 1)^2/(S_1 + S_2) = 0.48$ and $pv_a = \mathbf{P}\{\chi_1^2 > 0.48\} = 0.4884$. The data do not contradict the hypothesis.

**Comment 5.1.** In the case where $k = 2$, Friedman's test is equivalent to the sign test.

Indeed, $D_i > 0$ is equivalent to $R_{i1} = 2$, $R_{i2} = 1$, and $D_i < 0$ is equivalent to $R_{i1} = 1$, $R_{i2} = 2$, so

$$\bar{R}_{.1} = 1 + S_1/n, \quad \bar{R}_{.2} = 2 - S_1/n$$

and

$$S_F = \frac{12n}{6} 2(S_1/n - 1/2)^2 = \frac{(S_1 - n/2)^2}{n/4}$$

### 5.1.3. *Hypothesis on the median value*

Suppose that $X_1, ..., X_n$ is a simple sample of an absolutely continuous random variable $X$. Denote by $\theta$ the median of the random variable $X$.

**Hypothesis on the median value:** $H_0 : \theta = \theta_0$.

If the hypothesis $H_0$ holds then the differences $D_i = X_i - \theta_0$ are positive or negative with equal probabilities of 1/2. So the test statistic

$$S = \sum_{X_i - \theta_0 > 0} 1$$

is the number of positive differences $D_i$. The distribution of $S$ coincides with the distribution of the parametric sign test statistic $S_1$. So the exact and the asymptotic tests are defined as in section 5.1.1, replacing $S_1$ by $S$.

**Example 5.3.** (continuation of Example 4.8) Test the hypothesis in Example 4.8 that the median is a) greater than 15 $m^3$; b) equal to 15 $m^3$. The number of positive differences is $S_1 = 15$. The $P$-values are: a) $pv = I_{0.5}(n - S_1, S_1 + 1) = 0.0047$; b) $pv = 2I_{0.5}(n - S_1, S_1 + 1) = 0.0094$. Using the normal approximation with a continuity correction we have the asymptotic $P$-values a) $pv_a = \Phi(-2.5714) = 0.0051$; b) $pv_a = \mathbf{P}\{\chi_1^2 > 6.6122\} = 0.0101$. The hypotheses are rejected.

## 5.2. Runs test

**Definition 5.1.** *A* run *is a sequence of events of one type such that before and after this sequence an event of another type or no event occurs.*

Suppose that $m$ events $A$ and $n$ opposite events $\bar{A}$ occur during an experiment. These events can be arranged in a sequence in $C_N^m = C_N^n$ ways, $N = m + n$.

Denote by $V$ the number of runs in such a sequence. For example, in the sequence

$$A\,A\,\bar{A}\,A\,A\,\bar{A}\,\bar{A}\,A\,\bar{A}$$

there are $V = 6$ runs, $m = 5$, $n = 4$, $N = 9$.

**Randomness hypothesis for a sequence of two opposite events:** the probabilities of the appearance of all $C_N^m$ sequences are equal given $m$ and $n$.

Let us find the distribution of the number of runs $V$ under the randomness hypothesis.

**Theorem 5.1.** *Under the randomness hypothesis the distribution of the number of runs has the form*

$$\mathbf{P}\{V = 2i\} = \frac{2C_{m-1}^{i-1}C_{n-1}^{i-1}}{C_N^m}, \quad i = 1, ..., \min(m, n)$$

$$\mathbf{P}\{V = 2i + 1\} = \frac{C_{m-1}^{i-1}C_{n-1}^i + C_{m-1}^i C_{n-1}^{i-1}}{C_N^m}, \quad i = 1, ..., \min(m, n)$$

[5.3]

**Proof.** The number of all possible sequences with $m$ events $A$ and $n$ events $B$ is $C_N^m$.

Let us find the number of such sequences satisfying the equality $V = v$. Suppose that $v = 2i$ is even. Then there are $i$ runs of event $A$ and $i$ runs of event $\bar{A}$. In how many ways can a sequence of $m$ events $A$ be divided into $i$ subsequences? Imagine that screens are put between the symbols $A$. $i - 1$ are needed and there are $m - 1$ places for them. So a sequence from $m$ symbols $A$ can be divided into $i$ subsequences in $C_{m-1}^{i-1}$ ways. Analogously, the sequence of $n$ symbols $\bar{A}$ can be divided into $i$ subsequences in $C_{n-1}^{i-1}$ ways. So the number of elementary events favorable to the event $V = 2i$ is $2C_{m-1}^{i-1}C_{n-1}^{i-1}$. We multiply by 2 because the first run can begin with $A$ or $\bar{A}$. The classical definition of the probability implies

$$\mathbf{P}\{V = 2i\} = \frac{2C_{m-1}^{i-1}C_{n-1}^{i-1}}{C_N^m}, \quad i = 1, ..., \min(m, n)$$

The number of elementary events favorable to the event $V = 2i + 1$ is found analogously. There are two possibilities: $i + 1$ runs of $A$ and $i$ runs of $\bar{A}$, or $i$ runs of $A$ and $i + 1$ runs of $\bar{A}$. So

$$\mathbf{P}\{V = 2i + 1\} = \frac{C_{m-1}^{i-1}C_{n-1}^{i} + C_{m-1}^{i}C_{n-1}^{i-1}}{C_N^m}, \quad i = 1, ..., \min(m, n)$$

$\triangle$

The mean and the variance of the number of runs are

$$\mathbf{EV} = \frac{2mn}{N} + 1, \quad \mathbf{Var}(V) = \frac{2mn(2mn - N)}{N^2(N - 1)}$$

As $m, n \to \infty$, $m/n \to p \in (0, 1)$,

$$Z_{m,n} = \frac{V - \mathbf{EV}}{\sqrt{\mathbf{Var}(V)}} \xrightarrow{d} Z \sim N(0, 1) \qquad [5.4]$$

### 5.2.1. *Runs test for randomness of a sequence of two opposite events*

Runs can be used to test the above randomness hypothesis.

Suppose that the zero hypothesis is true. The appearance of sequences with long runs of the type

$$AAAAAAABBBBBBB, \quad ABBBBBBAAAAAAB$$

or with short runs of the type $ABABABABABABAB$ shows that there is some deterministic rule of arranging the objects, which contradicts the randomness hypothesis. So the zero hypothesis should be rejected if the number of runs $V$ is too small or too large.

**Runs test for randomness of a sequence of two opposite events:** the hypothesis $H_0$ is rejected with a significance level not greater than $\alpha$ if $V \leq c_1$ or $V \geq c_2$; here $c_1$ is the maximum integer satisfying the inequality $\mathbf{P}\{V \leq c_1|H_0\} \leq \alpha/2$, and $c_2$ is the minimum integer satisfying the inequality $\mathbf{P}\{V \geq c_2|H_0\} \leq \alpha/2$.

**Example 5.4.** Suppose that $k = 11$, and the events $A$ and $B$ occur $k_1 = 5$ and $k_2 = 6$ times, respectively. Let us find the $P$-value when the observed value $v$ of the number of runs $V$ is a) $v = 2$; b) $v = 3$; c) $v = 3$.

$$pv = 2\min\{F_V(v), 1-F_V(v-)\} = 2\min\{\mathbf{P}\{V \leq v\}, 1-\mathbf{P}\{V < v\}\}$$

By equation [5.3]

$$\mathbf{P}\{V = 2\} = 2\frac{C_4^0 C_5^0}{C_{11}^5} = \frac{2}{462} = 0.004329$$

$$\mathbf{P}\{V = 3\} = \frac{C_4^1 C_5^0 + C_4^0 C_5^1}{C_{11}^5} = \frac{9}{462} = 0.019481$$

$$\mathbf{P}\{V = 4\} = 2\frac{C_4^1 C_5^1}{C_{11}^5} = \frac{40}{462} = 0.086580$$

so

   a) $pv = 2\min\{0.004329, 1\} = 0.008658$;

   b) $pv = 2\min\{0.023810.0.9956716\} = 0.047619$;

   c) $pv = 2\min\{0.110390.0.952381\} = 0.220780$.

The same $P$-values are obtained if $V = 11$, 10, 9, respectively.

If $m$ and $n$ are large then the normal approximation [5.4] may be used.

**Asymptotic runs test:** the hypothesis $H_0$ is rejected with the approximate significance level $\alpha$ if $|Z_{k_1,k_2}| \geq z_{\alpha/2}$.

If $k$ is not large then a continuity correction is used:

**Asymptotic runs test with continuity correction:** the hypothesis $H_0$ is rejected with an approximate significance level $\alpha$ if

$$|Z^*_{k_1,k_2}| = |\frac{|V - \mathbf{E}V| - 0.5}{\sqrt{\mathbf{Var}(V)}}| \geq z_{\alpha/2}$$

The power of the test is poor if $m$ or $n$ is near to zero. This occurs if the probability of the event $A$ is very small or very near unity because, in such a case, the probability that the sequence of events contains events of the same type is large. The test is most powerful if $m$ is near $n$, which should occur if the probability of the event $A$ is near $0.5$.

Runs tests are used in more general situations.

**Example 5.5.** The event $A$ and the opposite event $B$ can occur during an experiment. $n = 40$ independent experiments were performed and the followings results were obtained:

$$ABBBBBBBABBBBABBBBBBBB$$

$$BABBBAABBAABABABAAAAAA$$

Test the hypothesis that As and Bs are arranged randomly. The number of runs is $V = 15$, $k_1 = 15$, $k_2 = 25$. Using the normal approximation with continuity correction we have $Z^*_{k_1,k_2} = 1.4549$ and $pv_a = 2(1 - \Phi(1.4549)) = 0.1457$. Using binomial distribution [5.6] (see section 5.2.3), we have $N = 34.185$, $p = 0.5192$ and an asymptotic $P$-value $0.1134 = 2I_{0.4808}(22, 14) < pv_a < I_{0.4808}(21, 14) = 0.1536$. Using linear interpolation with respect to $N$, we have $pv_a \approx 0.1462$. The data do not contradict the hypothesis.

### 5.2.2. Runs test for randomness of a sample

Suppose that a random vector $\mathbf{X} = (X_1, ..., X_n)^T$ is observed.

**Hypothesis of randomness of a sample** $H_0^*$: the sequence $\mathbf{X}$ is a simple sample, i.e. $X_1, ..., X_n$ are iid random variables.

The possible alternatives to the randomness hypothesis may be a trend or a cyclic pattern in the sample.

Set $D_i = X_i - \hat{M}$; here $\hat{M}$ is the median of the random variable $D_1$ under the hypothesis $H_0$ (if it is known under the hypothesis) or the sample median:

$$\hat{M} = \begin{cases} (X_{(\frac{n}{2}+1)} + X_{(\frac{n}{2})})/2, & \text{if } n \text{ is even,} \\ X_{(\frac{n+1}{2})}, & \text{if } n \text{ is odd;} \end{cases}$$

where $X_{(i)}$ is the $i$-th order statistic. Under the randomness hypothesis, the random variables $D_i = X_i - \hat{M}$ are equally distributed.

Exclude the elements of the sample satisfying the equality $D_i = 0$. Denote by $k_1$ and $k_2$ the numbers of events $A_i = \{D_i > 0\}$ and $B_i = \{D_i < 0\}$, respectively, and $k = k_1 + k_2$. Under the hypothesis $H_0^*$, the sequence of opposite events $A$ and $B$ satisfies the randomness hypothesis $H_0$, i.e. the hypothesis $H_0^*$ is narrower than the hypothesis $H_0$.

The above runs test can be used to test the hypothesis $H_0$. If the hypothesis $H_0$ is rejected then the hypothesis $H_0^*$ is also rejected.

**Example 5.6.** (continuation of Examples 4.3 and 4.4)

Test the randomness hypothesis using the data in Example 4.3 and the runs test.

The medians of the random variables $X_i$ are known: $\hat{M} = 0$. In this example $k_1 = 17$, $k_2 = 13$, $k = k_1 + k_2 = 30$.

The sequence of events $A$ and $B$ is

$$AAAAAABABABAABBBBBAAAAAAAABBBBBB$$

So

$$V = 8, \quad \mathbf{EV} = \frac{2 \cdot 17 \cdot 13}{30} + 1 = 15.733333$$

$$\mathbf{Var}V = \frac{2 \cdot 17 \cdot 13(2 \cdot 17 \cdot 13 - 30)}{30^2 \cdot 29} = 6.977165$$

The modified statistic with a continuity correction is

$$|Z^*_{k_1,k_2}| = |\frac{|V - \mathbf{EV}| - 0.5}{\sqrt{\mathbf{Var}V}}| = 2.738413$$

The asymptotic $P$ value is

$$pv_a = 2(1 - \Phi(|Z^*_{k_1,k_2}|)) = 2(1 - \Phi(2.738413)) = 0.006174$$

The randomness hypothesis is rejected with sufficiently small significance levels. The $P$-value of the runs test is considerably smaller than that of Kendall's and Spearman randomness tests but greater than that of Von Neuman's test. So, for this example, deviation from randomness was best seen by Von Neuman's, less well by the runs test, and worst by Kendall's and Spearman tests.

If we suppose that the median $\hat{M}$ is unknown and the sample median is used then

$$\hat{M} = (X_{(15)} + X_{(16)})/2 = (1 + 2)/2 = 1.5$$

and two positive differences become negative, so in such a case $k_1 = 15$, $k_2 = 15$, $k = 30$. The sequence of events $A$ and $B$ is

$$AAAAAABABAABBBBBAAAABBAABBBBBB$$

so $V = 10$. The statistic with continuity corrections is

$$|Z^*_{k_1 k_2}| = 2.043864, \quad \text{so} \quad pv_a = 0.040967$$

The randomness hypothesis is still rejected if the significance level 0.05 is chosen.

### 5.2.3. Wald–Wolfowitz test for homogeneity of two independent samples

Suppose that $\mathbf{X} = (X_1, ..., X_m)^T$ and $\mathbf{Y} = (Y_1, ..., Y_n)^T$ are two independent samples obtained by observing two independent absolutely continuous random variables $X \sim F$ and $Y \sim G$.

**Homogeneity hypothesis:** $H_0 : F(x) = G(x), \ \forall x \in \mathbf{R}$.

Omitting the indices in the *unified ordered sample*, we obtain a sequence of $m$ symbols $X$ and $n$ symbols $Y$. The Wald–Wolfowitz test is based on the number $V$ of runs in this sequence.

Under the homogeneity hypothesis, the symbols $X$ and $Y$ are arranged randomly, i.e. they are "mixed", so the number of runs should be large. Under the alternative $H_3 = H_1 \cup H_2$

$$\bar{H}_1 : \ F(x) \leq G(x), \ \exists x_0 : F(x_0) < G(x_0)$$

$$\bar{H}_2 : \ F(x) \geq G(x), \ \exists x_0 : F(x_0) > G(x_0)$$

Symbols of one type have a tendency to concentrate on one side, and symbols of another type on other side, of the unified ordered sample so the number of runs should be smaller under the alternative than under the hypothesis.

**Wald–Wolfowitz test:** the hypothesis $H_0$ is rejected with a significance level not greater than $\alpha$ if $V \leq k$; here $k$ – the maximum integer satisfying the equality

$$\mathbf{P}\{V \le k|H_0\} \le \alpha$$

If $V = v$ then the $P$-value $pv = F_V(v) = \mathbf{P}\{V \le v|H_0\}$ is obtained using formulas [5.3]. For small $m$ and $n$, the critical values can be computed by most statistical packages.

**Example 5.7.** Suppose that $m = 5$ and $n = 6$, $N = 5 + 6 = 11$. Let us find the $P$-value of the test when the observed value of the number of runs $V$ is a) $v = 2$; b) $v = 3$; c) $v = 4$.

By [5.3]

$$pv = F_V(v) = \mathbf{P}\{V \le v\}$$

$$\mathbf{P}\{V = 2\} = 2\frac{C_4^0 C_5^0}{C_{11}^5} = \frac{2}{462} = 0.004329$$

$$\mathbf{P}\{V = 3\} = \frac{C_4^1 C_5^0 + C_4^0 C_5^1}{C_{11}^5} = \frac{9}{462} = 0.019481$$

$$\mathbf{P}\{V = 4\} = 2\frac{C_4^1 C_5^1}{C_{11}^5} = \frac{40}{462} = 0.086580 \qquad [5.5]$$

so

a) $pv = 0.004329$; b) $pv = 0.023810$; c) $pv = 0.110390$.

If $m$ and $n$ are large then we use the [5.4] normal approximation:

**Asymptotic Wald–Wolfowitz test:** if $m$ and $n$ are large then the hypothesis $H_0$ is rejected with an approximate significance level $\alpha$ if $Z_{mn} \le -z_\alpha$.

For $N = m + n$ of medium size the normal approximation is more accurate if a continuity correction is used. Set

$$Z_{mn}^* = \begin{cases} \frac{V - \mathbf{E}V - 0.5}{\sqrt{\mathbf{Var}(V)}}, & \text{if } V - \mathbf{E}V > 0.5, \\ \frac{V - \mathbf{E}V + 0.5}{\sqrt{\mathbf{Var}(V)}}, & \text{if } V - \mathbf{E}V < -0.5, \\ 0, & \text{if } |V - \mathbf{E}V| \le 0.5. \end{cases}$$

**Asymptotic Wald–Wolfowitz test with continuity correction:** the hypothesis $H_0$ is rejected with an approximate significance level $\alpha$ if $Z^*_{mn} \le -z_\alpha$.

If $m$ and $n$ ($m < n$) are of medium size or if the ratio $m/n$ is small (see [BOL 83]) then the distribution of $V - 2$ is approximated by the binomial distribution $B(N, p)$ with

$$N = \frac{(m + n - 1)(2mn - m - n)}{m(m - 1) + n(n - 1)}, \quad p = 1 - \frac{2mn}{(m + n)(m + n - 1)}$$

[5.6]

**Asymptotic Wald–Wolfowitz test with binomial approximation:** the hypothesis $H_0$ is rejected if the number of runs $V$ is not greater than the critical values of the binomial distribution $B(N, p)$.

If $V = v$ then the approximate $P$-value is

$$pv_a = \sum_{i=0}^{v} C^i_{N+2} p^i (1 - p)^{N+2-i} = I_{1-p}(N + 2 - v, v + 1) =$$

$$1 - I_p(v + 1, N + 2 - v)$$

**Comment 5.2.**    Under the location alternative the power of the Wald–Wolfowitz test is smaller than the power of the Wilcoxon test, so the Wilcoxon test is recommended. Under the location-scale alternative, the Wald–Wolfowitz test is recommended.

**Example 5.8.** (continuation of Example 3.4) Use the data in Example 3.4 and the Wald–Wolfowitz test to test the hypothesis that fungicides have no influence on the percentage of infections.

We analyzed the data using the Kolmogorov–Smirnov and Wilcoxon–Mann–Whitney tests.

The sample sizes are $m = n = 7$, $N = m + n = 14$.

In the unified ordered sample, the observations of the first (A) and the second (B) groups are arranged in the following order

$$AAAAABBAABBBBB$$

The number of runs $V = 4$. The $P$-value computed using the SPSS package is $pv = 0.025058$.

The mean and the variance of the number of runs are

$$\mathbf{EV} = \frac{2mn}{m+n} + 1 = 8, \quad \mathbf{Var}(V) = \frac{2mn(2mn - m - n)}{(m+n)^2(m+n-1)} = 3.230769$$

Since $R - \mathbf{EV} = -4 < -0.5$, we use the approximation

$$Z_{m,n} = \frac{V - \mathbf{EV} + 0.5}{\sqrt{\mathbf{Var}(V)}} = -1.94722$$

The asymptotic $P$-value $pv_a = \Phi(-1.94722) = 0.025754$. The zero hypothesis $H_0$ is rejected with a significance level not less than $0.025058$.

For these data the deviation from the zero hypothesis was best seen using the Wilcoxon–Mann–Whitney test; the Wald–Wolfowitz test was less good, and the Kolmogorov–Smirnov test was the worst.

## 5.3. McNemar's test

Let us modify the sign test to the case of dependent samples.

Suppose that the marginal distributions of the random vector $(X_i, Y_i)^T$, $i = 1, \ldots, n$ are *Bernoulli*, i.e.

$$X_i \sim B(1, p_{1i}), \quad Y_i \sim B(1, p_{2i}), \quad p_{i1} = \mathbf{P}\{X_i = 1\}, \quad p_{i2} = \mathbf{P}\{Y_i = 1\}$$

where each random variable $X_i$ and $Y_i$ takes the value 1 if an event $A$ occurs and 0 otherwise.

**Homogeneity hypothesis:**

$$H_0 : \ p_{i1} = p_{i2} \quad \text{for all } i = 1, \ldots, n$$

The homogeneity hypothesis can be tested using the modified Friedman's test (with $k = 2$), which is based on the statistic $S_F^*$ (see equation [4.80]). We shall see that this test can be formulated in terms not only of the ranks but also in terms of the random variables $X_{ij}$. The modified Friedman's test written in such a form is called the McNemar's test. This test is parametric but there is a tradition in all statistical packages of attributing it to the group of non-parametric tests.

Let us consider typical situations where the formulated hypothesis is met.

**Example 5.9.** Suppose that before electoral debates $n$ people from various layers of society answer the question: "Will you vote for the candidate N?" After the electoral debates the same people are asked to answer the same question. Did the electoral debates change their opinion? The event $A$ is

$$A = \{\text{I shall vote for the candidate N}\}$$

$$p_{i1} = \mathbf{P}\{X_{i1} = 1\} \quad \text{and} \quad p_{i2} = \mathbf{P}\{X_{i2} = 1\}$$

are the probabilities that the $i$-th participant in the poll thinks of voting for the candidate $N$ before and after the debates, respectively.

**Example 5.10.** Investigation of the efficiency of headache remedies. $n$ people of various ages answered the question

"Does the remedy $j$ relieve headaches?"; $j = 1, 2$. Are the two remedies equally efficient? Then

$$A = \{\text{a remedy relieves the headache}\}$$

and $p_{i1}$ and $p_{i2}$ are the probabilities that the $i$-th participant in the survey thinks that the first or the second remedy, respectively, relieves headaches.

Let us consider the alternative

$$H_3 = H_1 \cup H_2$$

where

$H_1 : p_{i1} \leq p_{i2}$   for all $i = 1, \ldots, n$ $i_0$ exists such that $p_{i1} < p_{i2}$

$H_2 : p_{i1} \geq p_{i2}$   for all $i = 1, \ldots, n$ $i_0$ exists such that $p_{i1} > p_{i2}$

**Theorem 5.2.** *The hypothesis $H_0$ is equivalent to the hypothesis*

$$\mathbf{P}\{X_i = 1, Y_i = 0\} = \mathbf{P}\{X_i = 0, Y_i = 1\} \quad \text{for all } i = 1, \ldots, n$$

*The alternatives $H_1$ and $H_2$ are equivalent to the analogous hypotheses replacing the equalities by corresponding inequalities.*

**Proof.** Let us consider the conditional probabilities

$$\gamma_i = \mathbf{P}\{Y_i = 1 | X_i = 1\} \quad \text{and} \quad \beta_i = \mathbf{P}\{Y_i = 0 | X_i = 0\}$$

Then from the formula of complete probabilities

$$p_{i2} = \mathbf{P}\{X_i = 1\} = \gamma_i p_{i1} + (1 - \beta_i)(1 - p_{i1})$$

So the equality $p_{i1} = p_{i2}$ is equivalent to the equalities

$$(1 - \gamma_i)p_{i1} = (1 - \beta_i)(1 - p_{i1}) \Leftrightarrow \mathbf{P}\{Y_i = 0 | X_i = 1\}\mathbf{P}\{X_i = 1\} =$$

$$\mathbf{P}\{Y_i = 1|X_i = 0\}\mathbf{P}\{X_i = 0\} \Leftrightarrow \mathbf{P}\{X_i = 1, Y_i = 0\} = \mathbf{P}\{X_i = 0, Y_i = 1\}$$

The second part of the theorem is proved similarly by replacing equalities with inequalities.

$\triangle$

The theorem implies that it is sufficient to consider the objects with different results in the first and second experiments, i.e. such that the event

$$\{X_i = 1, Y_i = 0\} \cup \{X_i = 0, Y_i = 1\}$$

occurs. Under the zero hypothesis

$$\mathbf{P}\{X_i = 1, Y_i = 0 | \{X_i = 1, Y_i = 0\} \cup \{X_i = 0, Y_i = 1\}\} =$$

$$\mathbf{P}\{X_i = 0, Y_i = 1 | \{X_i = 1, Y_i = 0\} \cup \{X_i = 0, Y_i = 1\}\} = 0.5$$

So the test is constructed as follows: denote by $U_{kl}$ the number of objects satisfying

$$(X_i, Y_i) = (k, l), \ k, l = 0, \ 1; \quad U_{00} + U_{01} + U_{10} + U_{11} = n$$

| $k/l$ | 0 | 1 | |
|---|---|---|---|
| 0 | $U_{00}$ | $U_{01}$ | $U_{0.}$ |
| 1 | $U_{10}$ | $U_{11}$ | $U_{1.}$ |
| | $U_{.0}$ | $U_{.1}$ | $n$ |

**Table 5.1.** *The data*

By the theorem and the above discussion, only $m = U_{10} + U_{01}$ observations are used.

Under the hypothesis $H_0$, the conditional distribution of the statistic $U_{10}$, given $m = U_{10} + U_{01}$, is binomial $B(m, 1/2)$. So this conditional distribution is the same as that of the sign test statistic $S_1$ when $n = m$. We obtain the following.

**McNemar's test:** in the case of a two-sided alternative, the homogeneity hypothesis is rejected with a significance level not greater than $\alpha$ if

$$U_{10} \leq c_1 \quad \text{or} \quad U_{10} \geq c_2 \qquad [5.7]$$

where $c_1$ is the maximum integer satisfying the inequality

$$\mathbf{P}\{U_{10} \leq c_1\} = \sum_{k=0}^{c_1} C_m^k (1/2)^m = 1 - I_{0.5}(c_1 + 1, m - c_1) =$$

$$I_{0.5}(m - c_1, c_1 + 1) \leq \alpha/2$$

and $c_2$ is the minimum integer satisfying the inequality

$$\mathbf{P}\{U_{10} \geq c_2\} = \sum_{k=c_2}^{m} C_m^k (1/2)^m = I_{0.5}(c_2 . m - c_2 + 1) \leq \alpha/2$$

If $u$ is the observed value of $U_{10}$ then the $P$-value (see section 1.4) is

$$pv = 2 \min(F_{U_{10}}(u), 1 - F_{U_{10}}(u))$$

Using the results in section 5.1.1, we have

$$Q_2 = \frac{(U_{10} - U_{01})^2}{U_{10} + U_{01}} \xrightarrow{d} \chi^2(1) \quad \text{as} \quad n \to \infty$$

**Asymptotic McNemar's test:** if $n$ is large then the hypothesis $H_0$ is rejected with an approximate significance level $\alpha$ if

$$Q_2 > \chi_\alpha^2(1) \qquad [5.8]$$

For a medium sample size $n$, a continuity correction is used.

$$Q_2^* = \frac{(|U_{10} - U_{01}| - 1)^2}{U_{10} + U_{01}}$$

**Asymptotic McNemar's test with continuity correction:**
if $n$ is large then the hypothesis $H_0$ is rejected with an
approximate significance level $\alpha$ if

$$Q_2^* > \chi_\alpha^2(1) \qquad\qquad [5.9]$$

**Example 5.11.** (continuation of Example 5.9) Suppose that
before electoral debates 1000 people from various layers of
society answered the question: "Will you vote for the candidate
N?" After the electoral debates the same people were asked to
answer the same question. The results are given in the table
(the event $A$ is $A = \{$I shall vote for the candidate N$\}$, $B$ is
$B = \{$I shall not vote for the candidate N$\}$).

|   | A | B |   |
|---|---|---|---|
| A | 421 | 115 | 536 |
| B | 78 | 386 | 464 |
|   | 499 | 501 | 1000 |

**Table 5.2.** *The data*

Did the electoral debates change their opinion?

We have $n = V_{10} + V_{01} = 193$, $V_{10} = 78$, $pv = \mathbf{P}\{V_{10} \leq 78 | p =$
$0.5\} = 0.00938$. The hypothesis is rejected.

**Example 5.12.** Two different classifiers are tested with a
test set of size 500. The first classifier erroneously classifies
80 objects, the second misclassifies 60 objects. Classifiers are
often compared using the frequencies of errors. The classifiers
are tested with the same objects, so the observations are
dependent and such comparisons are incorrect. The McNemar
test should be used.

Suppose that in this example both classifiers correctly classified $U_{00} = 400$ objects and both erroneously classified $U_{11} = 40$ objects. Then the hypothesis that the second classifier is better is formulated as a parametric hypothesis $H_0 : p = 0.5$ with a one-sided alternative $\bar{H} : p < 0,5$ on the value of the parameter $p$ of the binomial distribution. The number of experiments is $n = U_{10} + U_{01} = 60$, and the number of appearances of the event is $U_{01} = 20$. The $P$-value is $pv = \mathbf{P}\{U_{01} \leq 20\} = 0.0067$. Using the normal approximation with a continuity correction we have $pv_a = 1 - \Phi(2.4529) = 0.0071$. We conclude that the second classifier is better.

**Example 5.13.** Ten experts verify units of two types and make one of the following decisions: "unit is good" (evaluation 1) or "unit is defective" (evaluation 0). Test the hypothesis that units of each type are of the same quality.

| Experts | 1 | 2 | 3 | 4 | 5 | 6 | 7 | 8 | 9 | 10 |
|---|---|---|---|---|---|---|---|---|---|---|
| First type | 1 | 1 | 0 | 0 | 0 | 0 | 1 | 0 | 1 | 0 |
| Second type | 1 | 1 | 1 | 0 | 1 | 1 | 0 | 0 | 1 | 1 |

We obtain:

$$U_{10} = 1, \quad U_{01} = 4, \quad U_{10} - U_{01} = -3, \quad m = U_{10} + U_{01} = 5$$

McNemar's test is equivalent to the test for the hypothesis $H_0 : p = 1/2$ on parameter values of the Bernoulli distribution in $m$ independent Bernoulli trials based on the number of successes $U_{10}$.

The value of the two-sided test statistic $U_{10}$ is 1 and the $P$-value (see section 1.4) is

$$pv = 2\min(F_{U_{10}}(1), 1 - F_{U_{10}}(0))$$

Since

$$F_{U_{10}}(0) = \frac{1}{2^5}C_5^0 = \frac{1}{32}, \quad F_{U_{10}}(1) = \frac{1}{2^5}(C_5^0 + C_5^1) = \frac{3}{16}$$

we have $pv = \frac{3}{16} = 0.1875$.

Using the asymptotic McNemar's test we obtain $Q = \frac{9}{5} = 1.8$. The approximate $P$-value (see section 1.4) is

$$pv_a = 1 - F_{\chi_1^2}(1.8) = 0.1797$$

Both tests conclude that the data do not contradict the hypothesis that units of both types have the same quality, because the $P$-values are not small.

## 5.4. Cochran test

Let us generalize the McNemar's test to the case of $k > 2$ random vectors.

Suppose that independent random vectors $(X_{i1}, \ldots, X_{ik})^T$, $i = 1, \ldots, n$ with Bernoulli marginal distributions, are observed

$$X_{i1} \sim B(1, p_{i1}), \ldots, X_{ik} \sim B(1, p_{ik})$$

$$p_{i1} = \mathbf{P}\{X_{i1} = 1\}, \ldots, p_{ik} = \mathbf{P}\{X_{ik} = 1\}$$

where any random variable $X_{i1}$ takes the value 1 if an event $A$ occurs and the value 0 if the event $A$ does not occur.

Write all observations $X_{ij}$ in Table 5.3.

In Table 5.3 we use the notation

$$X_{i.} = \sum_{j=1}^{k} X_{ij}, \quad i = 1, \ldots, n, \quad \text{and} \quad X_{.j} = \sum_{i=1}^{n} X_{ij}, \quad j = 1, \ldots, k$$

| $i \backslash j$ | 1 | 2 | ... | $k$ | $\Sigma$ |
|---|---|---|---|---|---|
| 1 | $X_{11}$ | $X_{12}$ | ... | $X_{1k}$ | $X_{1.}$ |
| 2 | $X_{21}$ | $X_{22}$ | ... | $X_{2k}$ | $X_{2.}$ |
| . | ... | ... | ... | ... | ... |
| $n$ | $X_{n1}$ | $X_{n2}$ | ... | $X_{nk}$ | $X_{n.}$ |
| $\Sigma$ | $X_{.1}$ | $X_{.2}$ | ... | $X_{.k}$ | |

**Table 5.3.** *The data*

**Homogeneity hypothesis:**

$$H_0: \; p_{i1} = ... = p_{ik}, \quad \text{for all } i = 1, \ldots, n \qquad [5.10]$$

Let us consider typical situations when such a hypothesis is tested.

**Example 5.14.** Suppose that using $k$ methods the presence of a virus is established in $n$ individuals of different ages. Applying the $j$-th method to the $i$-th individual one of two results is obtained: $X_{ij} = 1$ (the event $A = \{$virus is found$\}$ occurs) or $X_{ij} = 0$ (the event $\bar{A} = \{$virus is not found$\}$ occurs). Verify the hypothesis that all $k$ methods are equivalent. In this case, $p_{ij}$ is the probability that a virus is found in the $i$-th individual by the $j$-th method.

**Example 5.15.** Comparison of the efficiency of $k$ remedies for headache relief. $n$ people of various professions answered the question "Does the remedy $j$ relieve headaches?"; $j = 1, 2, ..., k$. If the $j$-th remedy ($j = 1, ..., k$) relieves the headache of the $i$-th individual then the random variable $X_{ij}$ takes the value 1, otherwise it takes the value 0. Test the hypothesis that the efficiencies of all $k$ remedies are the same. In this case, the event $A$ is $\{$remedy relieves headache$\}$, and $p_{ij}$ is the probability that the $j$-th remedy relieves the headache of the $i$-th individual.

**Example 5.16.** Ten experts examine units of four types and make one of the following decisions: "unit is good" (evaluation 1) or "unit is defective"(evaluation 0). Test the hypothesis that units of all four types are of the same quality.

The alternative

$$H_1 : \quad \text{there exist} \quad i, j, l : \ p_{ij} \neq p_{il}$$

If the distribution of $X_{ij}$ does not depend on the index $i$, i.e. $p_{ij} = p_j$ for all $i = 1, ..., n$, then the alternative is written in the form

$$H_1 : \quad \text{there exist} \quad j, l : p_j \neq p_l$$

The Cochran test is constructed as follows. Under the hypothesis $H_0$, the random variables $X_{.1}, ..., X_{.k}$ are equally distributed, so the values of the random variables $X_{.j} - \bar{X}$ are spread around zero; here

$$\bar{X} = \frac{1}{k} \sum_{j=1}^{k} X_{.j}$$

The spread of $X_{.j} - \bar{X}$ is characterized by

$$\sum_{j=1}^{k} (X_{.j} - \bar{X})^2 = \sum_{j=1}^{k} X_{.j}^2 - k\bar{X}^2$$

This statistic is proportional to the *Cochran* statistic

$$Q = \frac{k(k-1) \left( \sum_{j=1}^{k} X_{.j}^2 - k\bar{X}^2 \right)}{k \sum_{i=1}^{n} X_{i.} - \sum_{i=1}^{n} X_{i.}^2} \qquad [5.11]$$

**Comment 5.3.**    The Cochran statistic coincides with the modified Friedman statistic [4.80].

Indeed, $X_{ij}$ takes only two values: 1 or 0, so the $i$th line in Table 5.3 contains one (when $X_{i1} = \ldots = X_{ik}$) or two groups of coinciding observations.

If there is one group, i.e. $k_i = 1$, then its size is $t_{i1} = k$.

If there are two groups, i.e. $k_i = 2$, then their sizes are

$$t_{i1} = X_{i.}, \quad t_{i2} = k - X_{i.}$$

If $k_i = 1$ then $T_i = k^3 - k$. If $k_i = 2$ then

$$T_i = \sum_{j=1}^{2}(t_{ij}^3 - t_{ij}) = X_{i.}^3 + (k - X_{i.})^3 - (X_{i.} + (k - X_{i.})) =$$

$$k(3X_{i.}^2 - 3kX_{i.} + k^2 - 1)$$

In both cases, we can use the same formula $T_i = k(3X_{i.}^2 - 3kX_{i.} + k^2 - 1)$ because if $k_i = 1$ then $X_{i.} = 0$ or $X_{i.} = k$ , so

$$T_i = k(3 \cdot 0^2 - 3k \cdot 0 + k^2 - 1) = k^3 - k \quad \text{and}$$

$$T_i = k(3k^2 - 3kk + k^2 - 1) = k^3 - k$$

respectively. Hence

$$1 - \frac{1}{n(k^3 - k)}\sum_{i=1}^{n}T_i = \frac{3}{n(k^2 - 1)}(k\sum_{i=1}^{n}X_i - \sum_{i=1}^{n}X_i^2) \qquad [5.12]$$

Ordering the $i$-th line of Table 5.3 we obtain $k - X_{i.}$ zeros in the first $k - X_{i.}$ positions and $X_{i.}$ units in further positions.

If $X_{ij} = 0$ then

$$R_{ij} = \frac{1 + 2 + \cdots + (k - X_{i.})}{k - X_{i.}} = (k - X_{i.} + 1)/2$$

If $X_{ij} = 1$ then

$$R_{ij} = \frac{(k - X_{i.}) + (k - X_{i.} + 1) + \cdots + k}{X_{i.}} = k/2 + (k - X_{i.} + 1)/2$$

So

$$\sum_{j=1}^{k} R_{\cdot j}^2 = \sum_{j=1}^{k} (\frac{1}{2}\sum_{i=1}^{n}(k-X_{i\cdot}+1)+\frac{k}{2}X_{\cdot j})^2 = \sum_{j=1}^{k}(\frac{k}{2}(X_{\cdot j}-\bar{X})+n\frac{k+1}{2})^2 =$$

$$= \frac{k^2}{4}\sum_{j=1}^{k}(X_{\cdot j}-\bar{X})^2 + \frac{n^2 k(k+1)^2}{4}$$

We obtain

$$S_F = \frac{12}{k(k+1)n}\sum_{j=1}^{k} R_{\cdot j}^2 - 3n(k+1) = \frac{3k}{n(k+1)}\sum_{j=1}^{k}(X_{\cdot j}-\bar{X})^2$$

Using [4.80] and the expression for $S_F$ we obtain

$$S_F^* = \frac{S_R}{1 - \frac{1}{n(k^3-k)}\sum_{i=1}^{n} T_i} = \frac{k(k-1)\left(\sum_{j=1}^{k} X_{\cdot j}^2 - k\bar{X}^2\right)}{k\sum_{i=1}^{n} X_{i\cdot} - \sum_{i=1}^{n} X_{i\cdot}^2} = Q$$

**Cochran test:** the hypothesis $H_0$ is rejected with a significance level no greater than $\alpha$ if $Q \geq Q_\alpha$; here $Q_\alpha$ is the smallest real number $c$ verifying the inequality $\mathbf{P}\{Q \geq c|H_0\} \leq \alpha$.

The $P$-value is $pv = \mathbf{P}\{Q \geq q\}$; here $q$ is the observed value of the statistic $Q$.

If $n$ is large then by Theorem 4.15 the distribution of the Cochran statistic $Q = S_F^*$ is approximated by the chi-squared distribution with $k-1$ degrees of freedom.

**Asymptotic Cochran test:** if $n$ is large then the hypothesis $H_0$ is rejected with an approximate significance level $\alpha$ if

$$Q > \chi_\alpha^2(k-1) \tag{5.13}$$

**Example 5.17.** Ten experts verify units of four types and make one of the following decisions: "unit is good" (evaluation 1) or "unit is defective"(evaluation 0). Test the hypothesis that units of all four types are of the same quality. The results are given in Table 5.4.

| $i\,j$ | 1 2 3 4 | $X_{i.}$ |
|---|---|---|
| 1 | 1 1 1 0 | 3 |
| 2 | 1 1 1 0 | 3 |
| 3 | 0 1 0 1 | 2 |
| 4 | 0 0 1 1 | 2 |
| 5 | 1 1 0 0 | 2 |
| 6 | 1 1 1 0 | 3 |
| 7 | 1 1 1 0 | 3 |
| 8 | 0 0 0 1 | 1 |
| 9 | 1 1 0 0 | 2 |
| 10 | 0 1 0 1 | 2 |
| $X_{.j}$ | 6 8 5 4 | 23 |

**Table 5.4.** *The data*

In this example $k = 4$, $n = 10$

$$\sum_{i=1}^{10} X_{i.}^2 = 4\cdot 3^2 + 5\cdot 2^2 + 1^2 = 57, \quad \sum_{j=1}^{4} X_{.j}^2 = 6^2 + 8^2 + 5^2 + 4^2 = 141$$

$$\sum_{j=1}^{4} X_{.j} = \sum_{i=1}^{10} X_{i.} = 23, \quad \bar{X} = \frac{23}{4} = 5.75$$

$$Q = \frac{k(k-1)\left(\sum_{j=1}^{k} X_{.j}^2 - k\bar{X}^2\right)}{k\sum_{i=1}^{n} X_{i.} - \sum_{i=1}^{n} X_{i.}^2} = \frac{4\cdot 3(141 - 4\cdot 5,75^2)}{4\cdot 23 - 57} = 3$$

Using the SPSS package we obtain the $P$-value $pv = 0.466732$. The asymptotic $P$-value is $pv_a = \mathbf{P}\{\chi_3^2 > 3\} = 0.391625$. The data do not contradict the hypothesis that the quality of units of all types is the same.

**Comment 5.4.**    McNemar's test is a particular case of the Cochran test for $k = 2$.

Indeed, if $k = 2$ then

$$X_{i.} = \begin{cases} 0, & \text{if } X_{i1} = 0, X_{i2} = 0, \\ 1, & \text{if } X_{i1} = 1, X_{i2} = 0 \text{ or } X_{i1} = 0, X_{i2} = 1, \\ 2. & \text{if } X_{i1} = 1, X_{i2} = 1. \end{cases}$$

So the following relations between the data in Tables 5.1 and 5.2 hold

$$\sum_{i=1}^{n} X_{i.} = X_{1.} + X_{2.} = U_{10} + U_{01} + 2U_{11}, \quad \sum_{i=1}^{n} X_{i.}^2 = U_{10} + U_{01} + 4U_{11}$$

The denominator of [5.11] is

$$2\sum_{j=1}^{n} X_{j.} - \sum_{j=1}^{n} X_{j.}^2 = U_{10} + U_{01}$$

and the numerator is

$$X_{.1}^2 + X_{.2}^2 - 2\bar{X}^2 = X_{.1}^2 + X_{.2}^2 - \frac{1}{2}(X_{.1} + X_{.2})^2 =$$

$$\frac{(X_{.1} - X_{.2})^2}{2} = \frac{(U_{10} - U_{01})^2}{2}$$

Using expression [5.11] for the statistic $Q$ we obtain

$$Q = \frac{(U_{10} - U_{01})^2}{U_{10} + U_{01}}$$

i.e. the Cochran statistic coincides with the McNemar's statistic.

## 5.5. Special goodness-of-fit tests

Let $\mathbf{X} = (X_1, ..., X_n)^T$ be a simple sample of the rv $X$ with the cdf $F$ from a non-parametric class $\mathcal{F}$. We shall consider a composite hypothesis of the following form.

**Composite hypothesis:**

$$H_0 : F \in \mathcal{F}_0 = \{F_0(x, \boldsymbol{\theta}), \boldsymbol{\theta} \in \Theta \subset \mathbf{R}^s\} \subset \mathcal{F} \qquad [5.14]$$

where $F_0(x, \boldsymbol{\theta})$ is a cdf of specified form depending on an unknown parameter $\boldsymbol{\theta} \in \Theta$.

Using specific properties of the set $\mathcal{F}_0$, statistics with parameter-free probability distributions under $H_0$ can sometimes be found and tests may be based on such statistics.

Recall that such tests are the modified chi-squared, Cramér–von-Mises and Andersen–Darling tests (see sections 2.3 and 3.4) for specified location-scale (normal, logistic, Cauchy, extreme values distribution), and scale-shape (lognormal, Weibull, loglogistic distribution) families $\mathcal{F}_0$.

In this section we present other tests of this type.

### 5.5.1. *Normal distribution*

Suppose that $\mathcal{F}_0$ is the family of normal distributions $\{N(\mu, \sigma^2), -\infty < \mu < \infty, 0 < \sigma < \infty\}$.

**1. Tests based on the ratios of empirical moments.** The asymmetry and skewness coefficients of the normal random variable $X \sim N(\mu, \sigma^2)$ are

$$\gamma_1 = \mathbf{E}(X - \mu)^3/\sigma^3 = 0, \quad \gamma_2 = \mathbf{E}(X - \mu)^4/\sigma^4 - 3 = 0 \quad [5.15]$$

We shall also use the equality

$$\gamma_3 = \frac{\mathbf{E}|X - \mu|}{\sigma} = \sqrt{\frac{2}{\pi}}$$

Empirical analogs of $\gamma_1$ and $\gamma_2$ are the empirical asymmetry and skewness coefficients

$$g_1 = m_3/m_2^{3/2}, \quad g_2 = m_4/m_2^2 - 3, \quad m_k = \frac{1}{n}\sum_{j=1}^{n}(X_j - \bar{X})^k, \quad k = 2, 3, 4$$

[5.16]

and the empirical analog of $\gamma_3$ is

$$g_3 = \frac{1}{n}\sum_{j=1}^{n}|X_j - \bar{X}|/\sqrt{m_2}$$

[5.17]

The distribution of the random variable $Y = (X - \mu)/\sigma$, and consequently the distribution of the statistics $g_j$, does not depend on $\mu$ and $\sigma$. If $g_1, g_2$ or $g_3$ differ significantly from $\gamma_1 = 0$, $\gamma_2 = 0$ or $\gamma_3 = \sqrt{2/\pi}$, respectively, the zero hypothesis should be rejected.

More precisely, let us consider the wider hypotheses

$$H_1 : \gamma_1 = 0, \quad H_2 : \gamma_2 = 0, \quad H_3 : \gamma_3 = \sqrt{2/\pi}$$

**Tests based on the ratios of empirical moments:** the hypothesis $H_j$ is rejected if $g_j < c_{1j}$ or $g_j > c_{2j}$; here $c_{1j}$ and $c_{2j}$ are $(1 - \alpha/2)$ and $\alpha/2$, critical values of $g_j$ under $H_0$, $j = 1, 2, 3$.

The first two tests were considered by D'Agostino (see [DAG 70]); the third, introduced by Geary [GEA 35], is one of the oldest special tests for normality. The first two tests are called the *D'Agostino* $g_1$ and $g_2$ tests, the third is called *Geary's test*. For small $n$, the critical values of $g_j$ are tabulated (see [BOL 83]) and computed by most statistical software. They can also be found by simulation.

If $n$ is large then asymptotic tests are obtained which approximate the distribution of the statistics $g_j$ by the normal law.

The means and the variances of these statistics are

$$\mathbf{E}g_1 = 0, \quad \mathbf{E}g_2 = -\frac{6}{n+1}, \quad \mathbf{E}g_3 = \frac{2\Gamma((n+1)/2)}{\sqrt{\pi(n-1)}\Gamma(n/2)}$$

$$\mathbf{Var}(g_1) = \frac{6(n-2)}{(n+1)(n+3)}, \quad \mathbf{Var}(g_2) = \frac{24n(n-2)(n-3)}{(n+1)^2(n+3)(n+5)}$$

$$\mathbf{Var}(g_3) = \frac{1}{n}\left\{1 + \frac{2}{\pi}[\sqrt{n(n-2)} + \arcsin\frac{1}{n-1}]\right\} -$$

$$\frac{n-1}{\pi}\left[\frac{\Gamma((n+1)/2)}{\Gamma(n/2)}\right]^2$$

**Asymptotic test based on the ratios of empirical moments:** if $n$ is large then the hypothesis $H_j$ is rejected with an approximate significance level $\alpha$ if

$$|\bar{g}_j| = \frac{|g_j - \mathbf{E}g_j|}{\sqrt{\mathbf{Var}(g_j)}} > z_{\alpha/2}, \quad j = 1, 2, 3 \qquad [5.18]$$

**Comment 5.5.** The normality hypothesis is rejected if any of the hypotheses $H_1$, $H_2$ or $H_3$ is rejected. If all hypotheses are accepted then the distribution may be normal.

**Comment 5.6.** The convergence to the limit of the normal distribution is slow in the case of the statistic $g_2$.

**Example 5.18.** (continuation of Example 2.5). Using the data in Example 2.5 test the hypothesis that the obtained realizations of the empirical moments do not contradict the hypothesis that a random variable having a) normal; b) lognormal distribution was observed.

a) We have $\bar{g}_1 = 4.0697; \bar{g}_2 = 2.7696; \bar{g}_3 = -0.4213$. The respective asymptotic $P$-values are $4.71 \cdot 10^{-5}$, $0.0056$; $0.6735$. We conclude that the obtained realizations of empirical asymmetry and skewness coefficients contradict the normality of the random variable $V$.

b) After transformation, $Y_i = \ln X_i$, $i = 1, ..., 49$, we have $\bar{g}_1 = -1.8692; \bar{g}_2 = 0.2276; \bar{g}_3 = 0.8915$ and the respective asymptotic $P$-values: $0.0616$, $0.8200$; $0.3727$. We conclude that at the significance level $0.05$ the data do not contradict the lognormality hypothesis.

A modification of D'Agostino's $g_1$ and $g_2$ tests is the *Jarque–Bera* test: the test statistic is

$$T = n \left( \frac{\sqrt{b_1}}{6} + \frac{(b_2 - 3)^2}{24} \right)$$

Its distribution for large $n$ is approximated by the chi-squared distribution with two degrees of freedom. The hypothesis is rejected if $T$ takes large values.

**2. Sarkadi test.** Set

$$Y_j = X_{j+1} - \frac{1}{1 + \sqrt{n}} X_1 - \frac{n}{n + \sqrt{n}} \bar{X}, \quad j = 1, ..., n - 1$$

The random variables $Y_1, ..., Y_{n-1}$ (see exercise 5.9) are iid $N(0, \sigma^2)$. Define the rv

$$Z_j = \frac{Y_j}{\sqrt{\frac{1}{n-j-1}(Y_{j+1}^2 + ... + Y_{n-1}^2)}}, \quad j = 1, ..., n - 2$$

which are independent and have the Student distribution $Z_j \sim S(n - j - 1), j = 1, ..., n - 2$ (see exercise 5.10).

Denote by $S(x|\nu)$ the cdf of the Student distribution with $\nu$ degrees of freedom. The rv $U_j = S(Z_j|n - j - 1), j =$

$1, ..., n - 2$. are iid uniformly distributed in the interval $[0, \ 1]$, i.e. $U_1, ..., U_{n-2}$ is a simple sample of size $n - 2$ of the random variables $U \sim U(0, \ 1)$.

Let us consider the hypothesis $H_0^*$, stating that $(U_1, ..., U_{n-2})^T$ is a simple sample of the random variables $U \sim U(0, \ 1)$.

This hypothesis can be tested using any of the goodness-of-fit tests for simple hypotheses: Pearson chi-squared, Kolmogorov–Smirnov, Cramér–von-Mises, Andersen–Darling, etc.

If the hypothesis $H_0^*$ is rejected then the normality hypothesis is also rejected.

**Comment 5.7.** When defining the random variables $Y_j$, the random variable $X_1$ can be replaced by any $X_m$ with fixed $m$. Under location alternatives, it is natural to take $m = 1$ or $m = n$. If the alternatives suggest that at the moment $s$ the condition of the experiment might change then it is natural to take $m = s$.

**Example 5.19.** (continuation of Example 2.5). Using the data in Example 2.5, test the hypothesis that the obtained realizations of the empirical moments do not contradict the hypothesis that a random variable with a) normal; b) lognormal distribution was observed.

a) After the transform, we have $Z_1, ..., Z_{47}$, which, under the hypothesis, should be a simple sample of a random variable $Z \sim U(0, \ 1)$. To test the simple hypothesis $H_0^* : Z \sim U(0, \ 1)$ we use Pearson's chi-squared test and taking $k = 8$ intervals of equal probabilities we obtain the realization 14.9149 for the statistic $X_n^2$; the asymptotic $P$-value $pv_a = \mathbf{P}\{\chi_7^2 > 14.9149\} = 0.0107$. We conclude that the data contradict the hypothesis of normality of the random variable $V$. The values of the Kolmogorov–Smirnov, Cramer–Mizes and Andersen–Darling statistics are 0.1508, 0.2775 and 1.7396, the respective $P$-values are 0.2203, 0.1621 and 0.1299. These tests did not contradict the normality hypothesis.

b) Using the logarithms of the initial data, and after the transform, we have $Z_1, ..., Z_{47}$, which, under the the hypothesis, should be a simple sample of a random variable $Z \sim U(0, 1)$. To test the simple hypothesis $H_0^* : Z \sim U(0, 1)$ we use Pearson's chi-squared test and taking $k = 8$ intervals of equal probabilities we obtain the realization 1.8936 for the statistic $X_n^2$; the asymptotic $P$-value $pv_a = \mathbf{P}\{\chi_7^2 > 1,8936\} = 0.8637$. The values of the Kolmogorov–Smirnov, Cramer–Mizes and Andersen–Darling statistics are 0.1086, 0.0999 and 0.6934; the respective $P$ values are 0.6364, 0.5854 and 0.5644. The data do not contradict the lognormality hypothesis.

**3. Shapiro–Wilk test.** Suppose that $\mathbf{X} = (X_1, ..., X_n)^T$ is a simple sample from an absolutely continuous distribution with a mean $\mu = \mathbf{E}X_i$ and variance $\mathbf{Var}(X_i) = \sigma^2$.

Unbiased estimators of the parameters $\mu$ and $\sigma^2$ are

$$\bar{X}_n = \frac{1}{n} \sum_{i=1}^{n} X_i, \quad S_n^2 = \frac{1}{n-1} \sum_{i=1}^{n} (X_i - \bar{X}_n)^2$$

Under normal distribution, the standard deviation $\sigma$ can be estimated differently.

Denote by $\mathbf{X}^{(\cdot)} = (X_{(1)}, \ldots, X_{(n)})^T$ the vector of the order statistics.

Set $Z_{(i)} = (X_{(i)} - \mu)/\sigma$. Under the normality hypothesis, $Z_{(i)}$ is the $i$-th order statistic from the standard normal distribution and the distribution of the rv

$$\mathbf{Z}^{(\cdot)} = (Z_{(1)}, ..., Z_{(n)})^T$$

does not depend on the unknown parameters $\mu$ and $\sigma$ under the hypothesis $H_0$.

Denote by

$$\mathbf{m} = (m_{(1)}, ..., m_{(n)})^T \quad \text{and} \quad \mathbf{\Sigma} = [\sigma_{ij}]_{n \times n}$$

the mean and the covariance matrix of $\mathbf{Z}^{(\cdot)}$, which do not depend on unknown parameters $\mu$ and $\sigma$ under the hypothesis $H_0$.

Set

$$\boldsymbol{\theta} = (\mu, \sigma)^T, \quad \boldsymbol{C} = (\ 1 \mid \mathbf{m}\ ) = \begin{pmatrix} 1 & ... & 1 \\ m_{(1)} & ... & m_{(n)} \end{pmatrix}^T.$$

The equalities $X_{(i)} = \mu + \sigma Z_{(i)}$ imply that the mean and the covariance matrix of $\mathbf{X}^{(\cdot)}$ have the form:

$$\mathbf{EX}^{(\cdot)} = \boldsymbol{C}\boldsymbol{\theta}, \quad \mathbf{V} = \sigma^2 \mathbf{\Sigma}$$

so we have a linear model and the parameter $\boldsymbol{\theta}$ is estimated by the least-squares method

$$\hat{\boldsymbol{\theta}} = \boldsymbol{B}\mathbf{X}^{(\cdot)}, \quad \boldsymbol{B} = (\boldsymbol{C}^T \mathbf{\Sigma}^{-1} \boldsymbol{C})^{-1} \boldsymbol{C}^T \mathbf{\Sigma}^{-1} = \begin{pmatrix} c_1 & ... & c_n \\ a_1 & ... & a_n \end{pmatrix}.$$

In particular,

$$\hat{\sigma} = \sum_{i=1}^{n} a_i X_{(i)}, \quad \hat{\sigma}^2 = \left( \sum_{i=1}^{n} a_i X_{(i)} \right)^2$$

The vector $\mathbf{a} = (a_1, ..., a_n)^T$ has the form

$$\mathbf{a} = \frac{\mathbf{\Sigma}^{-1}\mathbf{m}}{\mathbf{m}^T \mathbf{\Sigma}^{-1} \mathbf{\Sigma}^{-1} \mathbf{m}}$$

so $\mathbf{a}^T \mathbf{a} = \sum_{i=1}^{n} a_i^2 = 1$.

Note that $BC = \mathbf{E}$, where $\mathbf{E}$ is a $2 \times 2$ unity matrix, so $\sum_{i=1}^{n} c_i = 1$, $\sum_{i=1}^{n} a_i = 0$.

The Shapiro–Wilk test is based on the statistic $W$ which is proportional to the ratio $\hat{\sigma}^2 / S_n^2$ of two consistent estimators of the parameter $\sigma^2$

$$W = \frac{\left( \sum_{i=1}^{n} a_i X_{(i)} \right)^2}{\sum_{i=1}^{n} (X_i - \bar{X}_n)^2}$$

The equalities $X_{(i)} = \mu + \sigma Y_{(i)}$ and the equality $\sum_{i=1}^{n} a_i = 0$ imply that the distribution of the statistic $W$ does not depend on unknown parameters under the hypothesis $H_0$.

The first estimator of the variance is good under the normality assumption. If the distribution is not normal, the first estimator may take values greater or smaller than in the normal case.

**Shapiro–Wilk test:** the normality hypothesis is rejected with a significance level not greater than $\alpha$ if $W < c_1$ or $W > c_2$, where $c_1$ is the maximum real number verifying the inequality $\mathbf{P}\{W < c_1 | H_0\} \leq \alpha/2$ and $c_2$ is the minimum real number verifying the inequality $\mathbf{P}\{W > c_2 | H_0\} \leq \alpha/2$.

Statistical software (for example SAS) computes the coefficients $a_i$ and the critical values of the statistic $W$.

**Example 5.20.** (continuation of Example 2.5). Using the data in Example 2.5, test the hypothesis that the obtained realizations of the empirical moments do not contradict the hypothesis that a random variable having a) normal and b) lognormal distribution was observed.

a) Using the SAS statistical software we obtain $W = 0.8706$ and the $P$-value $pv < 0.0001$. The hypothesis is rejected.

b) Using logarithms $\ln X_i$ and the SAS statistical software we have $W = 0.9608$ and the $P$-value $pv = 0.1020$. The data do not contradict the hypothesis of lognormality.

### 5.5.2. *Exponential distribution*

Let $X_1, ..., X_n$ be a simple sample of the absolutely continuous random variables $X$ with the cdf $F$ and the pdf $f$.

**Exponentiallity hypothesis:** $H_0 : X \sim \mathcal{E}(\lambda)$.

Under the zero hypothesis, the cdf of $X$ is $F(x) = 1 - e^{-\lambda x}, x \geq 0$.

We shall consider tests for the hypothesis $H_0$ using type two censored data when only first $r$-order statistics are observed. All these tests can be used in the case of complete simple samples, taking $r = n$ in all formulas.

The pdf of the vector $(X_{(1)}, X_{(2)}, ..., X_{(r)})^T$ of the first $r$-order statistics has the form

$$f_{X_{(1)},...,X_{(r)}}(x_1, \ldots, x_r) = \frac{n!}{(n-r)!}(1 - F(x_r))^{n-r} f(x_1), ..., f(x_r)$$

$$[5.19]$$

and is concentrated on the set $Q_r = \{(x_1, ..., x_r) : \quad -\infty < x_1 \leq \cdots \leq x_r < +\infty\}$.

In the particular case of the exponential distribution

$$f_{X_{(1)},...,X_{(r)}}(x_1, \ldots, x_r) = \frac{n!}{(n-r)!}\lambda^r e^{-\lambda(\sum\limits_{i=1}^{r} x_i + (n-r)x_r)}$$

$$0 \leq x_1 \leq \cdots \leq x_r < +\infty \qquad\qquad [5.20]$$

Set

$$Z_1 = nX_{(1)}, Z_2 = (n-1)(X_{(2)} - X_{(1)}), ..., Z_r =$$

$$(n - r + 1)(X_{(r)} - X_{(r-1)})$$

**Theorem 5.3.** *Under the hypothesis $H_0$, the random variables $Z_1, .., Z_r$ are iid and $Z_i \sim \mathcal{E}(\lambda)$.*

**Proof.** Using the change of variables

$$z_1 = nx_1, \ z_2 = (n-1)(x_2 - x_1), \ ... , \ z_r = (n - r + 1)(x_r - x_{r-1})$$

in pdf [5.20] we have

$$\sum_{i=1}^{r} x_i + (n-r)x_r = \sum_{i=1}^{r} z_i, x_1 = \frac{z_1}{n}, x_2 = \frac{z_2}{n-1}, ...,$$

$$x_r = \frac{z_1}{n} + \frac{z_2}{n-1} + ... + \frac{z_r}{n-r+1}$$

so the Jacobian is $(n-r)!/n!$. Putting the obtained expressions of $x_i$ into the pdf [5.5.2], and multiplying by the Jacobian we obtain the pdf of the random vector $(Z_1, .., Z_r)^T$:

$$f_{Z_1,..,Z_r}(z_1, \ldots, z_r) = \frac{n!}{(n-r)!} \lambda^r e^{-\lambda \sum\limits_{i=1}^{r} z_i}, \quad z_1, \ldots, z_r \geq 0 \quad [5.21]$$

$\triangle$

**1. Gnedenko test.** Set

$$G = \frac{r_2 \sum\limits_{i=1}^{r_1} z_i}{r_1 \sum\limits_{i=r_1+1}^{r} z_i}$$

where $r_1 = [r/2]$, $r_2 = r - [r/2]$.

**Theorem 5.4.** *Under the hypothesis $H_0$, the distribution of the random variable $G$ is Fisher, with $2r_1$ and $2r_2$ degrees of freedom.*

**Proof.** Theorem 5.3. implies that the random variables $2\lambda Z_1, ..., 2\lambda Z_r$ are iid with the cdf $1 - e^{-x/2}, x \geq 0$, i.e. each has a chi-squared distribution with two degrees of freedom. Hence, the random variables $\sum\limits_{i=1}^{r_1} z_i$ and $\sum\limits_{i=r_1+1}^{r} z_i$ are independent and have a chi-squared distribution with $2r_1$ and $2r_2$ degrees of freedom, respectively. This implies the result of the theorem.

$\triangle$

**Gnedenko test:** the hypothesis $H_0$ is rejected with a significance level $\alpha$ if $G < F_{1-\alpha/2}(2r_1, 2r_2)$ or $G > F_{\alpha/2}(2r_1, 2r_2)$.

The *P*-value is

$$pv = 2\min\{F_G(t), 1 - F_G(t)\}$$

where $F_G$ is the cdf of the random variable having a Fisher distribution with $2r_1$ and $2r_2$ degrees of freedom.

The Gnedenko test is powerful when the alternative is an increasing (or decreasing) hazard rate $\lambda(t)$ on the set $(0, \infty)$. If the hazard rate increases (decreases) then the statistic $G$ tends to take greater (smaller) values than in the case of an exponential distribution

**Example 5.21.** (continuation of Example 2.4). Test the hypothesis that the data in Example 2.4 are obtained by observing exponential random variables. The value of the statistic $G$ is $G = 2.1350$ and the respective $P$-value is $pv = 2\min(\mathbf{P}\{F_{35,35} < 2.1350\}, \mathbf{P}\{F_{35,35} > 2.1350\}) = 0.0277$. The hypothesis is rejected.

**2. Bolshev test.** Set

$$W_1 = \frac{Z_1}{Z_1 + Z_2}, \quad W_2 = \frac{Z_1 + Z_2}{Z_1 + Z_2 + Z_3}, \quad ...,$$

$$W_{r-1} = \frac{\sum\limits_{i=1}^{r-1} Z_i}{\sum\limits_{i=1}^{r} Z_i}, \quad W_r = \sum_{i=1}^{r} Z_i$$

$$U_i = W_i^i, \quad i = 1, ..., r-1.$$

**Theorem 5.5.** *Under the hypothesis $H_0$, the random variables $U_1, .., U_{r-1}$ are iid, having the uniform distribution $U(0,1)$.*

**Proof.** Change the variables in the pdf [5.21]:

$$w_1 = \frac{z_1}{z_1 + z_2}, \quad w_2 = \frac{z_1 + z_2}{z_1 + z_2 + z_3}, \quad ..., \quad w_{r-1} = \frac{\sum\limits_{i=1}^{r-1} z_i}{\sum\limits_{i=1}^{r} z_i}, w_r = \sum_{i=1}^{r} z_i$$

We have

$$z_1 = w_1 w_2 ... w_r, \quad z_2 = (1 - w_1) w_2 ... w_r, \quad z_{r-1} =$$

$$(1 - w_{r-2})w_{r-1}w_r, z_r = (1 - w_{r-1})w_r$$

The Jacobian is $w_2 w_3^2 ... w_{r-1}^{r-2} w_r^{r-1}$. Putting the obtained expressions for the variables $z_i$ into the pdf [5.21] and multiplying by the Jacobian, we obtain the pdf of the random vector $(W_1, .., W_r)^T$:

$$f_{W_1,...,W_r}(w_1, \ldots, w_r) = \frac{\lambda^r w_r^{r-1}}{(r-1)!} e^{-\lambda w_r} 1(2w_2)(3w_3^2)...(r-1)w_{r-1}^{r-2}$$

So the random variables $W_1, W_2 ...., W_{r-1}$ are independent, with the pdf

$$f_{W_i}(w_i) = i w_i^{i-1}, \quad 0 \le w_i \le 1, \quad i = 1, ..., r - 1$$

hence the result of the theorem.

$\triangle$

It can be shown (see [BOL 78]) that the formulated result is a characteristic property of an exponential distribution.

So, the hypothesis $H_0$ is equivalent to the hypothesis stating that $U_1, U_2 ...., U_{r-1}$ is the simple sample of size $r - 1$ from the uniform distribution $U(0, 1)$. Any of the goodness-of-fit tests such as Kolmogorov–Smirnov, Andersen–Darling or Cramér–von-Mises can be applied. For example, in the case of the Kolmogorov–Smirnov test, set $D_{r-1} = \max(D_{r-1}^+, D_{r-1}^-;$ here

$$D_{r-1}^+ = \max_{\le k \le r-1} (\frac{k}{r-1} - U_{(k)}), \quad D_{r-1}^- = \max_{\le k \le r-1} (U_{(k)} - \frac{k-1}{r-1})$$

and $U_{(k)}$ is the $k$-th order statistic, the statistic from the simple sample $U_1, U_2 ...., U_{r-1}$.

**Bolshev–Kolmogorov–Smirnov test:** the hypothesis $H_0$ is rejected with a significance level $\alpha$ if $D_{r-1} > D_\alpha(r - 1)$; here

$D_\alpha(r - 1)$ is the $\alpha$ critical value of the standard Kolmogorov distribution with $r - 1$ degrees of freedom.

*Bolshev–Andersen–Darling* and *Bolshev–Cramér–von-Mises* tests can be defined analogously.

**Example 5.22.** (continuation of Example 2.4). Test the hypothesis that the data in Example 2.4 are obtained by observing exponential random variables. After Bolshev's transform we obtain a sample $W_1, ..., W_{69}$, which, under the hypothesis, is a realization of a simple sample of a random variable $W \sim U(0, 1)$. We test this hypothesis using Pearson's chi-squared test with $k = 6$ equal probability intervals. The realization of the statistic $X_n^2$ is $X_n^2 = 11.6087$ and the $P$-value is $pv_a = \mathbf{P}\{\chi_5^2 > 11.6087\} = 0.0406$. The hypothesis is rejected. The realizations of the Kolmogorov–Smirnov, Cramér–von-Mises and Andersen–Darling statistics are 0.2212, 0.9829 it 4.8406, the respective $P$-values are 0.0025, 0.0034 and 0.0041. The hypothesis is rejected.

**3. Barnard test.** Set

$$V_1 = \frac{Z_1}{\sum\limits_{i=1}^{r} Z_i}, \quad V_2 = \frac{Z_1 + Z_2}{\sum\limits_{i=1}^{r} Z_i}, \quad ... \quad ,$$

$$V_{r-1} = \frac{\sum\limits_{i=1}^{r-1} Z_i}{\sum\limits_{i=1}^{r} Z_i}, \quad V_r = \sum\limits_{i=1}^{r} Z_i$$

**Theorem 5.6.** *Under the hypothesis $H_0$, the distribution of the random vector $(V_1, .., V_{r-1})^T$ coincides with the distribution of the vector of order statistics from a simple sample of size $r - 1$ with the elements having the distribution $U(0, 1)$.*

**Proof.** Change the variables in the pdf [5.21]:

$$v_1 = \frac{z_1}{\sum_{i=1}^{r} z_i}, \quad v_2 = \frac{z_1 + z_2}{\sum_{i=1}^{r} z_i}, \quad \ldots, \quad v_{r-1} = \frac{\sum_{i=1}^{r-1} z_i}{\sum_{i=1}^{r} z_i}, v_r = \sum_{i=1}^{r} z_i$$

We get

$$z_1 = v_1 v_r, \ z_2 = v_r(v_2 - v_1), \ z_{r-1} = v_r(v_{r-1} - v_{r-2}), z_r = v_r(1 - v_{r-1})$$

the Jacobian is $v_r^{r-1}$. Putting the obtained expression for the variables $z_i$ into the pdf [5.21] and multiplying by the Jacobian, we obtain the pdf of the random vector $(V_1, .., V_r)^T$:

$$f_{V_1,..,V_r}(v_1, \ldots, v_r) = \frac{\lambda^r v_r^{r-1}}{(r-1)!} e^{-\lambda v_r}(r-1)!,$$

$$0 \le v_1 \le \ldots \le v_{r-1}, v_r \ge 0$$

Hence the result of the theorem.

$\triangle$

The theorem implies that on replacing $U_{(k)}$ by $V_k$, $k = 1, \ldots, r-1$, in the statistic $D_{r-1}$ of the Bolshev–Kolmogorov–Smirnov test, the distribution of the obtained statistic $\tilde{D}_{r-1}$ does not change under the zero hypothesis. So the following test may be considered:

**Barnard–Kolmogorov–Smirnov test:** the hypothesis $H_0$ is rejected with a significance level $\alpha$ if $\tilde{D}_{r-1} > D_\alpha(r-1)$.

**Example 5.23.** (continuation of Example 2.4). Test the hypothesis that the data in Example 2.4 are obtained by observing exponential random variables. After Barnard's transform, we obtain the statistics $V_1, \ldots, V_{69}$, which, under the hypothesis, are the realizations of the order statistics from a uniform distribution $U(0, 1)$. We test this hypothesis using Pearson's chi-squared test with $k = 6$ equal probability intervals. The realization of the statistic $X_n^2$ is $X_n^2 = 13.0$ and the $P$-value is $pv_a = \mathbf{P}\{\chi_5^2 > 13.0\} = 0.0234$. The hypothesis is rejected. Realizations of the Kolmogorov–Smirnov, Cramer–Mizes and Andersen–Darling statistics are 0.2407, 1.3593 and 6.6310; the respective $P$-values are 0.0008, 0.0004 and 0.0007. The hypothesis is rejected.

### 5.5.3. *Weibull distribution*

Let $X_1, ..., X_n$ be a simple sample of an absolutely continuous rv $X$ with the cdf $F$ and the pdf $f$.

**Zero hypothesis** $H_0 : X \sim W(\theta, \nu)$.

The zero hypothesis states that the rv $X$ has the Weibull distribution with the cdf $F_X(x) = 1 - e^{-(x/\theta)^\nu}, x \geq 0$.

The hypothesis $H_0$ is equivalent the hypothesis that the rv $Y = \ln X$ have extreme value distributions with the cdf $F_Y(y) = 1 - e^{-e^{(x-\mu)/\sigma}}, x \in \mathbf{R}$; here $\mu = \ln \theta, \sigma = 1/\nu$.

The following test can be used when the data are type two censored, i.e. only first $r$-order statistics are observed. In the case of a complete simple sample, take $r = n$ in all formulas.

Let $(X_{(1)}, X_{(2)}, ..., X_{(r)})^T$ be the vector of the first $r$-order statistics. Set $Y_{(i)} = \ln X_{(i)}$, $r_1 = [r/2]$ and

$$S_{rn} = \frac{\sum\limits_{i=r_1+1}^{r-1} (Y_{(i+1)} - Y_{(i)})/E_{in}}{\sum\limits_{i=1}^{r-1} Y_{(i+1)} - Y_{(i)})/E_{in}}$$

where $Z_{(i)} = (Y_{(i)} - \mu)/\sigma$, $E_{in} = \mathbf{E}(Z_{(i+1)} - Z_{(i)})$. Using the equality $Y_{(i+1)} - Y_{(i)} = \sigma(Z_{(i+1)} - Z_{(i)})$, and the fact that the distribution of $Z_{(i)}$ does not depend on unknown parameters, we find that the means $E_{in}$ and the distributions of the statistic $S_{rn}$ also do not depend on unknown parameters. The Mann test is based on this statistic. Denote by $S_\alpha(r, n)$ the $\alpha$ critical value of the distribution of the statistic $S_{rn}$. The coefficients $E_{in}$ and the critical values $S_{rn,\alpha}$ are given by the SAS statistical software.

**Mann test:** the hypothesis $H_0$ is rejected with a significance level $\alpha$ if $S_{rn} < S_{1-\alpha/2}(r, n)$ or $S_{rn} > S_{\alpha/2}(r, n)$.

**Large samples.** If $n > 25$ then the distribution of the statistic $S_{rn}$ is approximated by the beta distribution with $r - r_1 - 1$ and $r_1$ degrees of freedom, and the coefficients $E_{in}$ are approximated by $1/[-(n - i)\ln(1 - i/n)]$, $i = 1, ..., r - 1$.

Denote by $\beta_{\alpha/2}(r - r_1 - 1, r_1)$ the $\alpha$ critical value of the beta distribution with $r - r_1 - 1$ and $r_1$ degrees of freedom.

**Asymptotic Mann test:** if $n$ is large then the hypothesis $H_0$ is rejected with an approximate significance level $\alpha$ if $S_{rn} < \beta_{1-\alpha/2}(r - r_1 - 1, r_1)$ or $S_{rn} > \beta_{\alpha/2}(r - r_1 - 1, r_1)$.

The Mann test is powerful against the lognormal alternative.

**Example 5.24.** (continuation of Example 2.4). In Examples 2.4, 5.20, 5.21 and 5.22 we saw that the data in Example 2.4 contradict the exponentiality hypothesis. Using the Mann test, let us test the hypothesis that these data are obtained by observing a random variable having the Weibull distribution. Using the SAS statistical software we have $S_{n,n} = 0.4294$ and the $P$-value is $pv = 0.2921$. The data do not contradict the hypothesis.


### 5.5.4. *Poisson distribution*

Suppose that the simple sample $X_1, ..., X_n$ of the discreet random variables $X$ take non-negative integer values.

**Zero hypothesis** $H_0$: $X \sim \mathcal{P}(\lambda), \lambda > 0$.

The zero hypothesis means that the distribution of the random variable $X$ is Poisson.

**1. Bolshev test.** Fix the sum $T_n = X_1 + ... + X_n$. Under the hypothesis $H_0$, the conditional distribution of the random vector $\mathbf{X} = (X_1, ..., X_n)^T$, given by $T_n = X_1 + ... + X_n$, is multinomial: $\mathbf{X} \sim \mathcal{P}_n(T_n, \boldsymbol{\pi}_0), \boldsymbol{\pi}_0 = (\pi_{i0}, ..., \pi_{n0})^T, \pi_{10} = ... = \pi_{n0} = 1/n$, i.e.

$$\mathbf{P}\{X_1 = x_1, ..., X_n = x_n | T_n\} = \frac{T_n!}{x_1!...x_n!} \left(\frac{1}{n}\right)^{T_n}$$

Bolshev [BOL 65] showed that the last equality is a characteristic property of a Poisson distribution. Using this fact, the wider hypothesis $H_0^*$ : $\boldsymbol{\pi} = \boldsymbol{\pi}_0$ on the equality of the multinomial probability vector $\boldsymbol{\pi}$ to a fixed vector $\boldsymbol{\pi}_0 = (1/n, ..., 1/n)^T$, given $T_n$ multinomial experiments, can be tested.

The hypothesis $H_0^*$ is tested using chi-squared statistics (see section 2.1):

$$X_n^2 = \sum_{j=1}^{n} \frac{(X_j - T_n/n)^2}{T_n/n} = \frac{n}{T_n} \sum_{j=1}^{n} X_j^2 - T_n$$

Under the hypothesis $H_0^*$, and for large $S_n$, the distribution of the statistic $X_n^2$ is approximated by a chi-squared distribution with $n - 1$ degrees of freedom.

**Bolshev test:** if $S_n$ is large then the hypothesi $H_0^*$ is rejected with an approximate significance level $\alpha$ if

$$X_n^2 < \chi_{1-\alpha/2}^2(n-1) \quad \text{or} \quad X_n^2 > \chi_{\alpha/2}^2(n-1) \qquad [5.22]$$

If the hypothesis $H_0^*$ is rejected then the zero hypothesis $H_0$ is also rejected.

The considered statistic is proportional to the ratio of unbiased and consistent estimators of the variance $\mathbf{Var}(X_i)$ and the mean $\mathbf{E}X_i$

$$X_n^2 = (n-1)\frac{s_n^2}{\bar{X}_n}, \quad \bar{X}_n = \frac{1}{n}\sum_{i=1}^{n} X_n, \quad s_n^2 = \frac{1}{n-1}\sum_{i=1}^{n}(X_i - \bar{X}_n)^2$$

If a specified alternative is considered then the test may be one-sided. Taking into consideration that $\mathbf{Var}(X_i)/\mathbf{E}X_i = 1$ for the Poisson distribution, $\mathbf{Var}(X_i)/\mathbf{E}X_i < 1$ for the binomial distribution and $\mathbf{Var}(X_i)/\mathbf{E}X_i > 1$ for the negative binomial distribution, and that the ratio $s_n^2/\bar{X}_n$ is an estimate of the ratio $\mathbf{Var}X_i/\mathbf{E}X_i$, it is natural to reject the zero hypothesis in favor of the *binomial alternative* if

$$X_n^2 < \chi_{1-\alpha}^2(n-1) \qquad [5.23]$$

and to reject it in favor of the *negative binomial alternative* if

$$X_n^2 > \chi_{\alpha}^2(n-1)$$

**Example 5.25.** Let us consider the results of the experiment conducted by Rutherford, Geiger and Chadwick (see [RUT 30]) in studies of $\alpha$-decay. The numbers of scintillations, $X_1, ..., X_n$, were registered on a zinc sulfide screen, as a result of penetration of the screen by $\alpha$ particles emitted from a radium source, in $n = 2,608$ mutually exclusive time intervals, each of duration 7.5 s. Let $n_j$ represent the frequency of the event $\{X = j\}$, i.e. $n_j$ is the number of values of $X_i$ equal to $j$. The frequencies $n_j$ from Rutherford's data are given in the following table.

| $j$ | 0 | 1 | 2 | 3 | 4 | 5 | 6 | 7 | 8 | 9 | 10 | 11 | 12 |
|---|---|---|---|---|---|---|---|---|---|---|---|---|---|
| $n_j$ | 57 | 203 | 383 | 525 | 532 | 408 | 273 | 139 | 45 | 27 | 10 | 4 | 2 |

The total number of scintillations registered is

$$T_n = \sum_{i=1}^{n} X_i = \sum_{j=1}^{\infty} j n_j = 10094$$

and the empirical moments are

$$\bar{X}_n = \frac{T_n}{n} = 3.8704, \quad s_n^2 = \frac{1}{n-1} \sum_{i=1}^{n} (X_i - \bar{X}_n)^2 =$$

$$\frac{1}{n-1} \sum_{j=0}^{\infty} n_j j^2 - n \bar{X}_n^2 = 3.6756$$

So $s_n^2 / \bar{X}_n = 0.9497 < 1$. Von Mises supposed that $X_1, ..., X_n$ follow the binomial distribution because the observed ratio is smaller than 1. Was von Mises correct?

Apply the chi-squared test [5.23]. For our data we obtain the value $X_n^2 = 2475.8$. The asymptotic $P$-value is

$$pv_a = \mathbf{P}\{\chi_{n-1}^2 < 2475.8\} = \mathbf{P}\{\chi_{2607}^2 < 2475.8\} = 0.03295$$

The zero hypothesis is rejected in favor of the binomial distribution if the significance level is greater than 0.03295.

Feller noted that the experimenters could not count the $\alpha$ particles themselves. In fact, they observed only scintillations caused by these particles. The $\alpha$ particles hitting the screen during the scintillation are not observed. During a time interval of length $\gamma > 0$ after the beginning of each scintillation, new $\alpha$ particles cannot be counted. Taking this phenomenon into account, the obtained conclusion does not contradict the exponential model of radioactive disintegration.

## 2. Empty boxes test.

If the parameter $\lambda$ of the Poisson distribution is small then most of the sample elements take the value 0 and the chi-squared approximation may not be exact. In such a case the "empty boxes" test may be used. This test is based on the statistic

$$Z_0 = \sum_{j=1}^{n} \mathbf{1}_{\{0\}}(X_j) \qquad [5.24]$$

i.e. the number of sample elements $X_j$ taking the value 0.

Suppose that there are $N$ balls and $n$ boxes. The balls are put into boxes in such a way that each ball, independently of the other balls, falls into any of the $n$ boxes with the same probability $1/n$. Denote by $Z_i = Z_i(n, N)$ the number of boxes containing exactly $i$ balls after the experiment. Then:

$$\sum_{i=0}^{N} Z_i = n, \quad \sum_{i=0}^{N} i Z_i = N$$

Let us find the distribution of the number $Z_0 = Z_0(n, N)$ of empty boxes.

**Theorem 5.7.** *Possible values of the random variable $Z_0 = Z_0(n, N)$ are* $\max(0, n - N) \leq k \leq n - 1$; *their probabilities are*

$$\mathbf{P}\{Z_0(n, N) = k\} = C_n^k (1 - \frac{k}{n})^N \mathbf{P}\{Z_0(n - k, N) = 0\} \qquad [5.25]$$

$$\mathbf{P}\{Z_0(n-k, N)\} = \sum_{l=0}^{n-k} C_{n-k}^l (-1)^l (1 - \frac{l}{n-k})^N$$

*The mean and the variance of the random variable $Z_0$ are*

$$\mathbf{E}Z_0(n, N) = n(1 - \frac{1}{n})^N$$

$$\mathbf{Var}\,Z_0(n, N) = n(n-1)(1-\frac{2}{n})^N + n(1-\frac{1}{n})^N - n^2(1-\frac{1}{n})^{2N} \quad [5.26]$$

**Proof.** Denote by $A_i$ the event signifying that the $i$-th box is empty, and the opposite events by $\bar{A}_i$. Then

$$\mathbf{P}\{Z_0(n, N) = k\} = \sum_{1 \le i_1 < ... < i_k \le n} \mathbf{P}\{A_{i_1} \cap ... \cap A_{i_k} \cap \bar{A}_{j_1} \cap ... \cap \bar{A}_{j_{n-k}}\}$$

where $\{j_1, ..., j_{n-k}\}$ is the complement of the set $\{i_1, ..., i_k\}$ to the set $\{1, 2..., n\}$. The number of members in the sum is $C_n^k$, and all probabilities are equal:

$$\mathbf{P}\{A_{i_1} \cap ... \cap A_{i_k}\}\mathbf{P}\{\cap \bar{A}_{j_1} \cap ... \cap \bar{A}_{j_{n-k}} | A_{i_1} \cap ... \cap A_{i_k}\} =$$

$$= (1 - \frac{k}{n})^N \mathbf{P}\{Z_0(n-k, N) = 0\}$$

The probability that the other $n - k$ boxes are not empty is

$$\mathbf{P}\{Z_0(n-k, N) = 0\} = 1 - \mathbf{P}\{Z_0(n-k, N) > 0\} = 1 - \mathbf{P}\{\cup_{i=1}^{n-k} A_i\} =$$

$$= 1 - \{\sum_i \mathbf{P}\{A_i\} - \sum_{i<j} \mathbf{P}\{A_i \cap A_j\} + \sum_{i<j<l} \mathbf{P}\{A_i \cap A_j \cap A_l\} - ...\} =$$

$$= 1 - (n-k)(1-\frac{1}{n-k})^N + C_{n-k}^2(1-\frac{2}{n-k})^N - C_{n-k}^3(1-\frac{3}{n-k})^N + ...$$

Formula [5.25] is proved.

Write the random variable $Z_0$ as a sum of equally distributed binomial random variables $Z_0 = Y_1 + Y_2 + ... + Y_n$;

the random variable $Y_i$ takes the value 1 if the $i$-th box is empty, and the value 0 otherwise. The equalities $\mathbf{P}\{Y_i = 1\} = (1 - \frac{1}{n})^N$ and $\mathbf{P}\{Y_i Y_j = 1\} = (1 - \frac{2}{n})^N, i \neq j$ imply

$$\mathbf{E}Z_0 = n\mathbf{E}Y_i = n(1 - \frac{1}{n})^N$$

$$\mathbf{Var}(Z_0) = n\mathbf{E}Y_i^2 + \sum_{i \neq j} \mathbf{E}(Y_i Y_j) - (\mathbf{E}Z_0)^2 =$$

$$= n(n-1)(1 - \frac{2}{n})^N + n(1 - \frac{1}{n})^N - n^2(1 - \frac{1}{n})^{2N}$$

$\triangle$

Under the zero hypothesis $H_0$ the number $Z_0(n, S_n)$ of sample elements taking the value 0 has the distribution [5.26] if we replace $N$ by $S_n$. If the hypothesis is not true then the the balls fall into the boxes with different probabilities, so the random variable $Z_0(n, S_n)$ tends to take greater values.

**Empty boxes test:** the hypothesis is rejected with a significance level not greater than $\alpha$ if

$$Z_0(n, S_n) \geq c_n \qquad [5.27]$$

where $c_n$ is the smallest integer satisfying the inequality

$$\mathbf{P}\{Z_0(n, S_n) \geq c_n\} \leq \alpha$$

If $n$ is large an approximate test is obtained, approximating $Z_0(n, S_n)$ by the normal distribution.

**Asymptotic empty boxes test:** if $n$ is large then the hypothesis is rejected with an approximate significance level $\alpha$ if

$$\frac{Z_0 - \mathbf{E}Z_0}{\sqrt{\mathbf{Var}(Z_0)}} > z_\alpha. \qquad [5.28]$$

**Example 5.26.** Some chromosomes exposed to $X$-rays mutate. $n = 793$ chromosomes were exposed to $X$-rays. The numbers $n_i$ of chromosomes with $i$ mutations are given in the following table.

| $i$ | 0 | 1 | 2 | 3 | $\Sigma$ |
|---|---|---|---|---|---|
| $n_i$ | 639 | 141 | 13 | 0 | 793 |

Test the hypothesis that the number $X$ of mutations has the Poisson distribution $X \sim \mathcal{P}(\lambda)$. We have $Z_0(n, S_n) = 639$, $S_n = \sum_i i n_i = 167$, and $\mathbf{E}Z_0 = 642.327$, $\mathrm{Var}Z_0 = 12.3516$. Using the normal approximation, we have the asymptotic $P$-value $pv_a = 1 - \Phi(0.9466) = 0.1722$. The data do not contradict the hypothesis.

**Comment 5.7.**   The empty boxes test is also used to test the simple hypothesis $H_0 : F(x) \equiv F_0(x)$. Divide the abscissas axes into $k$ intervals $-\infty = a_0 < a_1 < ... < a_k = +\infty$ in such a way that $F_0(a_j) - F_0(a_{j-1}) = 1/k$, $j = 1, ..., k$. Under the hypothesis, $n$ elements of the sample (balls) are randomly put into $k$ intervals (boxes). The number $Z_0(k, n)$ of intervals not containing sample elements (empty boxes) has distribution [5.26], taking $k$ instead of $n$ and $n$ instead of $S_n$. The hypothesis $H_0$ is rejected using tests [5.27] or [5.28].

## 5.6. Bibliographic notes

Runs tests were introduced by Wald and Wolfowitz [WAL 40]. McNemar's and Cochran's test were given by McNemar [MCN] and Cochran [COC 54].

Special normality tests have been considered by many authors: Geary [GEA 35], Kac *et al.* [KAC 55], Shapiro and Wilk [SHA 65], D'Agostino [DAG 70, DAG 71, DAG 90], Shapiro and Francia [SHA 72], Bowman and Shenton

[BOW 75], Hegazy and Green [HEG 75], Locke and Spurrier [LOC 76], Prescott [PRE 76], Spiegelhalter [SPI 77], Frosini [FRO 78], Lin and Mudholkar [LIN 80], Sarkadi [SAR 81, FRO 87], Epps and Pulley [EPP 83], Best and Reiner [BES 85], Jarque and Bera [JAR 87], Park [PAR 99], and Zhang [ZHA 99].

Baringhaus *et al.* [BAR 89], Lemeshko *et al.* [LEM 05, LEM 09a, LEM 09b] conducted a comparative analysis of a number of statistical tests for testing for deviations from the normal distribution. The powers were analyzed and shortcomings of particular tests were revealed that had not been previously mentioned in the literature.

Special tests for exponential distributions have been considered by Gnedenko *et al.* [GNE 65], Bolshev [BOL 66, BOL 87] and others. For extensive surveys of such tests, see Ascher [ASC 90], and Balakrishnan and Basu [BAL 95].

Special tests for the Weibull distribution have been considered by Mann *et al.* [MAN 73], Cabana and Quiroz [CAB 05], and others. For extensive surveys on such tests see [RIN 09].

Special tests for the Poisson distribution have been considered by Bolshev [BOL 65], Kolchin *et al.* [KOL 67], and Baringhaus and Henze [BAR 92].

## 5.7. Exercises

**5.1.** Apply the sign test to test the hypothesis on the equivalence of methods of starch quantity determination using the data in exercise 4.3.

**5.2.** Apply the sign test to test the hypothesis on the equal efficiency of both seeders using the in exercise 4.4.

**5.3.** Apply the runs test to test the hypothesis on the equality of poison effects in two experiments (see exercise 3.14).

**5.4.** Show that the ASE of the sign test with respect to the Student's test for the hypothesis $H : \theta = \theta_0$ ($\theta$ is the median of a symmetric continuous distribution with the cdf $f(x)$) against location alternatives is $4\sigma^2(f(\theta_0))^2$.

**5.5.** Using the data in exercise 2.16 and the runs test verify the randomness hypothesis.

**5.6.** Using the data in exercise 2.17 and the runs test verify the randomness hypothesis.

**5.7.** In a public opinion poll the same 3000 electors were interrogated before elections and after elections on their opinion on a parliamentary party. Before the elections 300 electors had negative opinions. A year after the elections 350 electors had negative opinions; 150 electors did not change their negative opinion, 150 electors changed their opinion from negative to positive and 200 changed from positive to negative. Did the rating of the party change?

**5.8.** (continuation of **5.7.**). Suppose that the same 3,000 electors were interrogated before the next elections. Two hundred and seventy electors had negative opinions. Of these 180 had positive opinions in the first two polls and 70 had positive opinions in one of the preceding polls and negative opinions in the other. Did the rating of the party change?

**5.9.** Let $X_1, ..., X_n$ be a simple sample of $X \sim N(\mu, \sigma^2)$ and $Y_i = X_{i+1} - X_1/(1 + \sqrt{n}) - n\bar{X}/(n + \sqrt{n})$. Prove that $Y_1, ..., Y_{n-1}$ are iid random variables, $Y_i \sim N(0, \sigma^2)$.

**5.10.** (continuation of exercise 5.9). Prove that $Z_1, ..., Z_{n-2}$ are independent random variables having a Student distribution, $Z_i \sim S(n - i - 1)$, if $Z_j = Y_j \sqrt{n - j - 1}/(Y_{j+1}^2 + ... + Y_{n-1}^2)$.

**5.11.** Using the data in exercise 2.16 and the tests in section 5.5.1, verify the normality hypothesis.

**5.12.** Using the data in exercise 2.17 and the tests in section 5.5.1, verify the lognormality hypothesis.

**5.13.** Using the data in exercise 2.18, apply the empty boxes test.

## 5.8. Answers

**5.1.** Among 13 non-zero differences $S = 3$ are negative and $pv = 2\mathbf{P}\{S \leq 3\} = 0.0923$. The hypothesis is rejected if the significance level is greater than 0.0923.

**5.2.** Among nine non-zero differences $S = 1$ are negative and $pv = 2\mathbf{P}\{S \leq 1\} = 0.0195$. The hypothesis is rejected if the significance level is greater than 0.0195.

**5.3.** The number of runs is $V = 29$. The $P$-value $\mathbf{P}\{V \geq 29\} = 1.2 \cdot 10^{-6}$, so by the Wald–Wolfovitz runs test the data do not contradict the hypothesis.

**5.5.** The number of runs is $V = 54, k_1 = 50, k_2 = 50$. We have $Z_{k_1,k_2}^* = 0.5025$ and $pv_a = 2(1 - \Phi(0.5025)) = 0.6153$. The data do not contradict the hypothesis.

**5.6.** The number of runs $V = 90, k_1 = 76, k_2 = 74$. We have $Z_{k_1,k_2}^* = 2.4906$ and $pv_a = 2(1 - \Phi(2.4906)) = 0.0128$. The hypothesis is rejected.

**5.7.** Apply McNemar's test. The number of electors whose opinion changed from positive to negative is 200, and from negative to positive the number is 150. The hypothesis $H$ : $p = 1/2$ on the value of binomial probability is tested given $U_{01} + U_{10} = n = 350$ binomial experiments with the number of successes $U_{10} = 150$. We have $pv = 2\min(\mathbf{P}\{U_{10} \leq 150\}, \mathbf{P}\{U_{10} \geq 150\}) = 0.0087$. The hypothesis is rejected.

**5.8.** The data are sufficient to count the value of the Cochran statistic. We have $Q = 14.8485$ and $pv_a = \mathbf{P}\{\chi_2^2 > 14.8485\} = 0.0006$. The hypothesis is rejected.

**5.11.** The values of the test statistics are $\bar{g}_1 = 2.1598, \bar{g}_2 = 0.1524, \bar{g}_3 = -1.0849$ and the respective $P$-values are 0.0308, 0.8788 and 0.2779. The value of the empirical asymmetry coefficient contradicts the normality hypothesis if the significance level is greater than 0.0308. Applying the Sarkadi transform to the elements of the data, we obtain a simple sample $Z_1, ..., Z_{n-2}$ from the uniform distribution $U(0,1)$. Applying the Pearson chi-squared test to test the hypothesis $H : Z \sim U(0, 1)$ and choosing $k = 8$ intervals of equal probabilities we have $X^2 = 5.5102$ and $pv_a = 0.5980$. Applying the Kolmogorov–Smirnov, Cramér–von-Mises and Andersen–Darling tests we have the following values as the test statistics: 0.1071, 0.2629 and 1.6185. The respective $P$-values are 0.2039, 0.1800 and 0.1487. The value of the Shapiro–Wilk statistic is equal to 0.9716 and the respective $P$-value is 0.0293. The Shapiro–Wilk test rejects the normality hypothesis. Other tests did not detect the contradictions to the hypothesis.

**5.12.** Using the logarithms of the data elements, the values of the statistics $\bar{g}_1, \bar{g}_2, \bar{g}_3$ are 0.5901, 0.8785 and 0.3765 and the respective $P$-values are 0.5551, 0.3797 and 0.7065. Tests based on empirical moments do not detect contradictions to the normality hypothesis. Applying the Sarkadi transform, we have a simple sample $Z_1, ..., Z_{n-2}$ from the uniform distribution $U(0,1)$). Applying the Pearson chi-squared test to test the hypothesis $H : Z \sim U(0, 1)$, and choosing $k = 10$ intervals of equal probabilities we have $X^2 = 6.1892$ and $pv_a = 0.7208$. Applying the Kolmogorov–Smirnov, Cramér–von-Mises and Andersen–Darling tests we have the following values of the test statistics: 0.0548, 0.0752 and 0.5594. The respective $P$-values are $> 0.25$. The value of the Shapiro–Wilk statistic is 0.9920 and the respective $P$-value is 0.5644. None of the tests rejects the normality hypothesis.

**5.13.** a) The numbers of empty boxes $Z_0^{(i)}$, $i = 1, 2, 3, 4$, are 2870, 593, 639, 359. Find $\mathbf{E}(Z_0^{(i)})$, $\mathrm{Var}(Z_0^{(i)})$ and $\bar{Z}_0^{(i)}$ using formula [5.27]. The values of $\bar{Z}_0^{(i)}$ are 0.399; 0.937; $-0.947$; $-0.826$. The respective $P$-values are 0.345; 0.174; 0.828; 0.796. There is no reason to reject the hypothesis. b) The number of empty boxes $Z_0$ in the unified sample is 1871. We find $\mathbf{E}Z_0$, $\mathrm{Var}(Z_0)$, and $\bar{Z}_0 = -0.107$; $pv_a = 0.543$. There is no reason to reject the hypothesis.

# APPENDICES

# Appendix A

# Parametric Maximum Likelihood Estimators: Complete Samples

Let
$$\mathbf{X} = (X_1, \ldots, X_n)^T$$
be a simple sample, i.e. $X_1, \ldots, X_n$ are independent random variables. Suppose that $X_i \sim p(x, \boldsymbol{\theta}), \boldsymbol{\theta} \in \Theta \subset \mathbf{R}^m$, where $p(x, \boldsymbol{\theta})$ is the pdf of $X_i$ with respect to a $\sigma$-finite measure $\mu$. The pdf of $\mathbf{X}$ is $f(\boldsymbol{x}, \boldsymbol{\theta}) = \prod_{i=1}^n p(\boldsymbol{x}_i, \boldsymbol{\theta})$.

The likelihood function and its logarithm are

$$L(\boldsymbol{\theta}) = \prod_{i=1}^n p(X_i, \boldsymbol{\theta}), \quad \ell(\boldsymbol{\theta}) = \ln L(\boldsymbol{\theta})$$

respectively. The Fisher information matrix is

$$\boldsymbol{I}(\boldsymbol{\theta}) = \mathbf{E}_{\boldsymbol{\theta}} \dot{\ell}(\boldsymbol{\theta}) \dot{\ell}(^T\boldsymbol{\theta}) = -\mathbf{E}_{\boldsymbol{\theta}} \ddot{\ell}(\boldsymbol{\theta}) = n i(\boldsymbol{\theta})$$

$$\boldsymbol{i}(\boldsymbol{\theta}) = -\mathbf{E}_{\boldsymbol{\theta}} \ddot{\ell}_1(\boldsymbol{\theta}), \qquad \ell_1(\boldsymbol{\theta}) = \ell_1(\boldsymbol{\theta}, X_1) = \ln p(X_1, \boldsymbol{\theta})$$

Denote by $\mathbf{P}_{\boldsymbol{\theta}}$ the probability measure defined by the model: $\mathbf{P}_{\boldsymbol{\theta}}\{X_1 \in A\} = \int_A p(x, \boldsymbol{\theta}) d\mu(x)$ for any Borel set $A$.

We consider *identifiable* models: if $\theta_1 \neq \theta_2$ then $\mathbf{P}_{\theta_1} \neq \mathbf{P}_{\theta_1}$, i.e. there exists a Borel set $A$: $\mathbf{P}_{\theta_1}\{X_1 \in A\} \neq \mathbf{P}_{\theta_2}\{X_1 \in A\}$.

For any quadratic matrix $A = (a_{ij})_{n \times n}$, consider the norm
$$\| A \| = (\sum_{i=1}^{n} \sum_{j=1}^{n} a_{ij}^2)^{1/2}.$$

**Conditions A:** Suppose that:

1) the set $\Theta$ is open;

2) for almost all $y \in \mathbf{R}$ in the neighborhood $V_\rho = \{\theta : \|\theta - \theta_0\| \leq \rho\}$ of the true value $\theta_0$ of the parameter $\theta$ there exist continuous derivatives

$$\dot{p}(y, \theta) = (\frac{\partial}{\partial \theta_1} p(y, \theta), \ldots, \frac{\partial}{\partial \theta_m} p(y, \theta))^T$$

$$\ddot{p}(y, \theta) = \left[ \frac{\partial^2}{\partial \theta_i \partial \theta_j} p(y, \theta) \right]_{m \times m}$$

3) in the neighborhood $V_\rho$

$$\int_{\mathbf{R}} \dot{p}(y, \theta) \mu(dy) = \frac{\partial}{\partial \theta} \int_{\mathbf{R}} p(y, \theta) \mu(dy) = \mathbf{0}$$

$$\int_{\mathbf{R}} \ddot{p}(y, \theta) \mu(dy) = \frac{\partial}{\partial \theta} \int_{\mathbf{R}} \dot{p}(y, \theta) \mu(dy) = \mathbf{0}$$

4) the Fisher information matrix $i(\theta)$ is positive definite;

5) there exist non-negative functions $h$ and $b$ such that for almost all $y \in \mathbf{R}$ and all $\theta \in V_\rho$

$$\| \ddot{\ell}_1(y, \theta) - \ddot{\ell}_1(y, \theta_0) \| \leq h(y) b(\theta), \quad \mathbf{E}_{\theta_0}\{h(\mathbf{X}_1)\} < \infty, \quad b(\theta_0) = 0$$

and the function $b$ is continuous at the point $\theta_0$.

**Theorem A.1.** *Under conditions A there exists a sequence of random variables* $\{\hat{\boldsymbol{\theta}}_n\}$ *such that*

$$\mathbf{P}(\dot{\ell}(\hat{\boldsymbol{\theta}}_n) = 0) \to 1, \quad \hat{\boldsymbol{\theta}}_n \xrightarrow{P} \boldsymbol{\theta}_0, \tag{A.1}$$

$$\sqrt{n}(\hat{\boldsymbol{\theta}} - \boldsymbol{\theta}_0) = i^{-1}(\boldsymbol{\theta}_0) \frac{1}{\sqrt{n}} \dot{\ell}(\boldsymbol{\theta}_0) + o_P(1). \tag{A.2}$$

$$\sqrt{n}(\hat{\boldsymbol{\theta}} - \boldsymbol{\theta}_0) \xrightarrow{d} N_m(0, i^{-1}(\boldsymbol{\theta}_0)). \tag{A.3}$$

$$\frac{1}{\sqrt{n}} \dot{\ell}(\boldsymbol{\theta}_0) \xrightarrow{d} N_m(0, i(\boldsymbol{\theta}_0)), \tag{A.4}$$

$$-\frac{1}{n} \ddot{\ell}(\boldsymbol{\theta}_0) \xrightarrow{P} i(\boldsymbol{\theta}_0), \tag{A.5}$$

$$-\frac{1}{n} \ddot{\ell}(\hat{\boldsymbol{\theta}}) \xrightarrow{P} i(\boldsymbol{\theta}_0) \tag{A.6}$$

**Comment A.1.** In general the sequence of random variables $\hat{\boldsymbol{\theta}}_n$ depends on $\boldsymbol{\theta}_0$, so these variables are not necessarily estimators. Nevertheless, if the equations $\dot{\ell}(\boldsymbol{\theta}) = 0$ have solutions for every $n$ then usually a consistent sequence of solutions of these equations can be found using the theorem. Indeed, if for every $n$ there is unique solution $\tilde{\boldsymbol{\theta}}_n$ then $\hat{\boldsymbol{\theta}}_n = \tilde{\boldsymbol{\theta}}_n$. If the number of solutions is greater than one and there is a consistent sequence of estimators $\bar{\boldsymbol{\theta}}_n$ (for example estimators obtained by the method of moments) then the sequence of solutions nearest to $\bar{\boldsymbol{\theta}}_n$ may be used.

**Comment A.2.** Under the conditions A

$$-(\hat{\boldsymbol{\theta}}_n - \boldsymbol{\theta}_0)^T \ddot{\ell}(\hat{\boldsymbol{\theta}}_n)(\hat{\boldsymbol{\theta}}_n - \boldsymbol{\theta}_0) \xrightarrow{d} \chi_m^2, \quad \dot{\ell}^T(\boldsymbol{\theta}_0) i^{-1}(\boldsymbol{\theta}_0) \dot{\ell}(\boldsymbol{\theta}_0) \xrightarrow{d} \chi_m^2, \tag{A.7}$$

$$-\dot{\ell}^T(\boldsymbol{\theta}_0) \ddot{\ell}^{-1}(\boldsymbol{\theta}_0) \dot{\ell}(\boldsymbol{\theta}_0) \xrightarrow{d} \chi_m^2, \quad -\dot{\ell}^T(\boldsymbol{\theta}_0) \ddot{\ell}^{-1}(\hat{\boldsymbol{\theta}}_n) \dot{\ell}(\boldsymbol{\theta}_0) \xrightarrow{d} \chi_m^2. \tag{A.8}$$

**Comment A.3.** Under the conditions A

$$-2\ln\frac{L(\boldsymbol{\theta}_0)}{L(\hat{\boldsymbol{\theta}}_n)} \xrightarrow{d} \chi^2(m) \qquad\qquad\qquad \text{[A.9]}$$

Set

$$\Theta_0 = \{\boldsymbol{\theta} : \boldsymbol{\theta} = \varphi(\boldsymbol{\lambda}),\ \boldsymbol{\lambda} \in G\}, \quad G \subset \mathbf{R}^{m-k}, \quad k < m$$

here $\varphi : G \to \Theta$ is a continuously differentiable map.

**Comment A.4.** Under conditions A

$$R = -2\ln\frac{\sup_{\boldsymbol{\theta}\in\Theta_0} L(\boldsymbol{\theta})}{\sup_{\boldsymbol{\theta}\in\Theta} L(\boldsymbol{\theta})} \xrightarrow{d} \chi^2(k),\ n \to \infty$$

# Appendix B

# Notions from the Theory of Stochastic Processes

## B.1. Stochastic process

A finite set $\mathbf{X} = (X_1, \ldots, X_k)^T$ of random variables defined on the same probabilistic space $(\Omega, \mathcal{F}, \mathbf{P})$ is called a random vector. It induces probability measure $\mathbf{P_X}$ in $(\mathbf{R}^k, \mathcal{B}^k)$

$$\mathbf{P_X}(A) = \mathbf{P}\{\omega : \mathbf{X}(\omega) \in A\}, \quad A \in \mathcal{B}^k$$

where $\mathcal{B}^k$ is the $\sigma$ algebra of Borel sets in $\mathbf{R}^k$.

The notion "stochastic process" generalizes the notion "random vector" to the case where the number of random variables defined on the same probabilistic space $(\Omega, \mathcal{F}, \mathbf{P})$ is infinite (not necessary countable).

Suppose that $\mathcal{T}$ is a subset of the real line.

**Definition B.1.** A system of random variables $\{X(t, \omega), t \in \mathcal{T}, \omega \in \Omega\}$ defined on the same probability space $(\Omega, \mathcal{F}, \mathbf{P})$ is called a *stochastic process*.

In the particular case of a $k$-dimensional random vector the set $\mathcal{T}$ is $\{1, 2, \ldots, k\}$. For any fixed elementary event $\omega \in \Omega$, a non-random function $x(t) = X(t, \omega)$ is defined on the set $\mathcal{T}$. This function is called the *trajectory* or the *path* of the stochastic process $X$.

Denote by $D = \{X(\cdot, \omega), \omega \in \Omega\}$ the space of all trajectories. A stochastic process can be treated as a random function taking values in the space of the trajectories. $\sigma$-Algebras of subsets of the space of trajectories may be constructed in various ways.

Let $\rho(x, y)$ be the distance between functions $x$ and $y$ from $D$. In this book it suffices to consider the supremum norm distance

$$\rho(x, y) = \sup_{\in} \mid x(t) - y(t) \mid \qquad \text{[B.1]}$$

The set $G \subset D$ is open if with any $x \in G$ there exists an open ball $B_\varepsilon(x) = \{y : \rho(x, y) < \varepsilon\} \subset G$.

**Definition B.2.** *The smallest $\sigma$-algebra containing all open sets of $D$ is called the Borel $\sigma$-algebra of $D$ and is denoted by $\mathcal{B}(D)$.*

**Definition B.3.** *The probability distribution of the stochastic process $\{X(t), t \in \mathcal{T}\}$ is the probability measure in the space $(D, \mathcal{B}(D))$ for any $A \in \mathcal{B}(\mathcal{X})$ satisfying the equality*

$$\mathbf{P}^X(A) = \mathbf{P}\{X \in A\}$$

## B.2. Examples of stochastic processes

### B.2.1. *Empirical process*

Suppose that $\mathbf{X} = (X_1, \ldots, X_n)^T$ is a simple sample of a random variable $X$ with cumulative distribution function

$F(t) = \mathbf{P}\{X \le t\}$ and

$$\hat{F}_n(t) = \frac{1}{n} \sum_{i=1}^{n} \mathbf{1}_{(-\infty,t]}(X_i)$$

is the empirical distribution function.

**Definition B.4.** *The stochastic process*

$$\mathcal{E}_n(t) = \sqrt{n}(\hat{F}_n(t) - F(t)), \ t \in \mathcal{T} = \mathbf{R} \qquad \text{[B.2]}$$

*is called the empirical process.*

If $F(t)$ is absolutely continuous then the empirical process $\mathcal{E}_n$ can be transformed to the stochastic process

$$\mathcal{E}_n^*(y) = \sqrt{n}(\hat{G}_n(y) - y), \ y \in [0,1] \qquad \text{[B.3]}$$

where $\hat{G}_n(y)$ is the empirical distribution function of the simple sample of size $n$ of the random variable $Y = F(X) \sim U(0,1)$. The stochastic process $\mathcal{E}_n^*(t)$ takes values from the interval $[-\sqrt{n}, \sqrt{n}]$ when $y \in [0,1]$. Its trajectories are right continuous functions with finite left limits, with jumps of size $1/\sqrt{n}$. By Theorem 3.1

$$\mathbf{E}(\mathcal{E}_n^*(y)) = 0, \quad \mathbf{Cov}(\mathcal{E}_n^*(y), \mathcal{E}_n^*(z)) = y(1-z), \quad 0 \le y \le z \le 1 \qquad \text{[B.4]}$$

### B.2.2. *Gauss process*

*A stochastic process is called a Gauss process if all its finite-dimensional distributions are normal.*

### B.2.3. *Wiener process (Brownian motion)*

**Definition B.5.** *A stochastic process $W(t), t \in \mathcal{T} = [0, \infty)$ is called the Wiener process (Brownian motion) if it verifies the following conditions:*

*a)* $W(0) = 0$;

*b)* $W(t) - W(s) \sim N(0, t - s)$ *for all* $0 \leq s < t < \infty$;

*c)* $W$ *has independent increments, i.e. for any* $k \in \mathbf{N}$ *and* $0 = t_0 < t_1 < \cdots < t_k$, *the random variables* $W(t_{j+1}) - W(t_j)$, $j = 1, 2, \ldots, k - 1$, *are independent.*

Assumptions a)–c) uniquely define finite-dimensional distributions of the Wiener process

$$(W(t_1), \ldots, W(t_n))^T \sim N_n(\mathbf{0}, \boldsymbol{\Sigma}), \quad \boldsymbol{\Sigma} = [\sigma_{ij}]_{n \times n}, \quad \sigma_{ij} = t_i \wedge t_j$$

It can be shown that finite-dimensional distributions uniquely define Wiener process probability distributions.

### B.2.4. *Brownian bridge*

**Definition B.6.** *The stochastic process*

$$B(t) = W(t) - tW(1), \quad t \in [0, 1] \qquad \text{[B.5]}$$

*is called a Brownian bridge on the interval* $[0, 1]$.

Finite-dimensional distributions are normal: for all natural $n$ and real $0 \leq t_1 < \cdots < t_n \leq 1$

$$(B(t_1), \ldots, B(t_n))^T \sim N_n(\mathbf{0}, \boldsymbol{\Gamma})$$

$$\boldsymbol{\Gamma} = \|\gamma_{ij}\|_{n \times n}, \quad \gamma_{ij} = t_i(1 - t_j), 0 \leq t_i \leq t_j \leq 1 \qquad \text{[B.6]}$$

Note that the means and the covariance matrices of finite-dimensional distributions of the Brownian bridge and of the empirical process $\mathcal{E}_n^*(t)$ given in section B.21 coincide.

Brownian motion and Brownian bridge are Gauss processes.

## B.3. Weak convergence of stochastic processes

Suppose that a sequence of stochastic processes $\{X^{(n)}\}$ and a stochastic process $X$ are defined in the same probability space $(\Omega, \mathcal{F}, \mathbf{P})$. Denote by $\mathbf{P}^{X^{(n)}}$ and $\mathbf{P}^X$ the probability distributions of these processes in the measurable space $(D, \mathcal{B}(D))$.

Let us consider the space $D$, where $D = D(\mathcal{T})$ is a space of cadlag (right-continuous and having finite left limits) functions on an interval of the real line $\mathcal{T}$. Denote by $D^p = D_1 \times \cdots \times D_p$ the product of $p$ such spaces.

Let $\rho(x, y)$ be the distance between the functions $x$ and $y$, $x, y \in D$. It can be defined in various ways. In the space of cadlag functions $D = D[0, \tau]$, the most used are Skorokhod and supremum norm distances. Let $\Lambda = \{\lambda(\cdot)\}$ be the set of strictly increasing continuous functions on $[0, \tau]$ such that $\lambda(0) = 0, \lambda(\tau) = \tau$.

*Skorokhod metric*

$$\rho(x, y) =$$

$$\inf\{\varepsilon > 0 : \text{exists } \lambda \in \Lambda : \sup_{0 \leq t \leq \tau} \mid \lambda(t) - t \mid \leq \varepsilon,$$

$$\sup_{0 \leq t \leq \tau} \mid x(t) - y(\lambda(t)) \mid \leq \varepsilon\} \qquad \text{[B.7]}$$

*Supremum norm metric*

$$\rho(x, y) = \sup_{0 \leq t \leq \tau} \mid x(t) - y(t) \mid \qquad \text{[B.8]}$$

The smallest $\sigma$-algebra containing all open sets of $D$ is called the Borel $\sigma$-algebra of $D$ and is denoted by $\mathcal{B}(D)$. The smallest $\sigma$-algebra containing all sets of the form $A = A_1 \times \cdots A_s$, where $A_1, \cdots, A_s$ are open sets of $D_1, \cdots, D_s$, respectively, is called the *Borel $\sigma$-algebra* of $D^s$ and is denoted by $\mathcal{B}^s(D^s)$.

Denote by $\partial A$ the border of the set $A \in \mathcal{B}$.

**Definition B.7.** A sequence $\{X^{(n)}\}$ weakly converges to $X$ if for any $A \in \mathcal{B}^s(D)$, such that $\mathbf{P}^X(\partial A) = 0$, the following convergence holds

$$\mathbf{P}^{X^{(n)}}(A) \to \mathbf{P}^X(A), \quad n \to \infty$$

As in the case of random variables and random vectors, weak convergence is denoted by $X^{(n)} \xrightarrow{d} X$.

Weak convergence implies that for all $x_1, \ldots, x_m$

$$(X^{(n)}(x_1), \ldots, X^{(n)}(x_m)) \xrightarrow{d} (X(x_1), \ldots, X(x_m))$$

but not necessarily *vice versa*.

If the limit process has continuous trajectories (this is the case in all developments in this book) then convergence using Skorokhod and supremum norm metrics is equivalent.

## B.4. Weak invariance of empirical processes

Suppose that $\mathbf{X} = (X_1, \ldots, X_n)^T$ is a simple sample of a random variable $X$ with the cdf $F$, and $\hat{F}_n$ is the empirical distribution function.

By the central limit theorem for sums of random vectors, the finite-dimensional vectors $(\mathcal{E}_n(x_1), \ldots, \mathcal{E}_n(x_m))^T$ of the empirical process

$$\mathcal{E}_n(x) = \sqrt{n}(\hat{F}_n(x) - F(x))$$

weakly converge to the random vector

$$(Z_1, \ldots, Z_m)^T \sim N_m(\mathbf{0}, \mathbf{\Gamma}), \quad \mathbf{\Gamma} = [\gamma_{ij}]_{m \times m}$$

$$\gamma_{ij} = F(x_i)(1 - F(x_j)), \quad i \leq j = 1, \ldots, m,$$

for any $m$ and $-\infty < x_1 < \cdots < x_m < \infty$.

This result and the central limit theorem imply that the finite-dimensional distributions of an empirical process weakly converge to the finite-dimensional distributions of the stochastic process $B(F(x))$, where $B$ is the Brownian bridge.

More generally, if $h$ is a continuous function then

$$(h(\mathcal{E}_n(x_1)), \ldots, h(\mathcal{E}_n(x_m))) \xrightarrow{d} (h(B(F(x_1))), \ldots, h(B(F(x_m))))$$
[B.9]

A more general theorem holds:

**Theorem B.1.** (Weak invariance of the empirical process). *If $F$ is an absolutely continuous cumulative distribution function and $h$ is a continuous functional on the set of right continuous functions with finite left limits and defined on the interval $[0, 1]$ then*

$$h(\mathcal{E}_n^*) \xrightarrow{d} h(B)$$

*where*

$$\mathcal{E}_n^*(y) = \sqrt{n}(\hat{G}_n(y) - y), \quad \hat{G}_n(y) = \frac{1}{n} \sum_{i=1}^{n} \mathbf{1}_{(-\infty, y]}(F(X_i)),$$

*$B$ is the Brownian bridge on $[0, 1]$*

$\triangle$

## B.5. Properties of Brownian motion and Brownian bridge

**Property 1.** If $\tau = \inf\{t : W(t) \in A\}$, $A$ is a Borel set of real numbers on the line and $W(t)$ is a Brownian motion then $\tilde{W}(t) = W(t + \tau) - W(\tau)$ is also a Brownian motion.

**Proof.** We have: $\tilde{W}(0) = 0$ and for all $0 = t_0 < t_1 < \cdots < t_k$, $x_1, \ldots, x_k$

$$\mathbf{P}\{\tilde{W}(t_1) - \tilde{W}(t_0) \leq x_1, \ldots, \tilde{W}(t_k) - \tilde{W}(t_{k-1}) \leq x_k\} =$$

$$\int_0^\infty \mathbf{P}\{W(t_1 + u) - W(u) \leq x_1, \ldots,$$

$$W(t_k + u) - W(t_{k-1} + u) \leq x_k | \tau = u\} dF_\tau(u) =$$

$$\mathbf{P}\{W(t_1 + u) - W(u) \leq x_1\} \ldots \mathbf{P}\{W(t_k + u) - W(t_{k-1} + u) \leq x_k\},$$

because the event $\{\tau = u\}$ is defined by the stochastic process $W(s), s \leq u$, the conditional probability under the integral sign coincides with the unconditional probability.

$\triangle$

**Property 2.** (Reflexion rule) *For all* $x, y \in \mathbf{R}, t \geq t_0 \geq 0$

$$\mathbf{P}\{W(t) > x + y | W(t_0) = x\} = \mathbf{P}\{W(t) < x - y | W(t_0) = x\}$$

**Proof.** We have

$$\mathbf{P}\{W(t) > x + y | W(t_0) = x\} =$$

$$\mathbf{P}\{W(t_0) + (W(t) - W(t_0)) > x + y | W(t_0) = x\} =$$

$$\mathbf{P}\{W(t) - W(t_0) > y\} = 1 - \Phi(y/\sqrt{t - t_0})$$

Analogously

$$\mathbf{P}\{W(t) < x - y | W(t_0) = x\} = \mathbf{P}\{W(t) - W(t_0) < -y\} =$$

$$1 - \Phi(y/\sqrt{t - t_0})$$

$\triangle$

**Property 3.** *For any* $x \in \mathbf{R}$

$$P(x) = \mathbf{P}\{\exists t \in [0, 1] : B(t) = x\} = e^{-2x^2}$$

**Proof.** $P(0) = 1$, $P(-x) = P(x)$ because $B$ and $-B$ have the same probability distributions. So it suffices to consider the case $x > 0$.

Let us consider the stochastic process $W(t) = B(t) + tW_1$, where $W_1 \sim N(0,1)$ is a random variable which does not depend on $B$. Then $W(t), t \in [0,1]$, is the Brownian motion because it is a Gaussian process

$$\mathbf{E}W(t) = 0, \quad \mathbf{Cov}(W(s), W(t)) = s \wedge t.$$

Since $B(1) = 0$, we have $W(1) = W_1$

Let us consider the conditional probability

$$P(x, \varepsilon) = \mathbf{P}\{\exists t \in [0,1] : W(t) \geq x \,|\, |W(1)| < \varepsilon\}, \quad 0 < \varepsilon < x$$

Set $\tau = \inf\{t : W(t) > x\}$. Since the trajectories of $W$ are continuous, $W(\tau) = x$. Property 1 implies that

$$\tilde{W}(u) = W(\tau + u) - W(\tau)$$

is Brownian motion. If $\tau < 1$ then

$$W(1) = W(\tau + (1 - \tau)) = \tilde{W}(1 - \tau) + x$$

Hence, using the fact that $\tilde{W}$ and $-\tilde{W}$ have identical probability distributions, we obtain

$$P(x, \varepsilon) = \mathbf{P}\{\tau < 1 \,|\, |W(1)| < \varepsilon\} =$$

$$\mathbf{P}\{\tau < 1, |\tilde{W}(1 - \tau) + x| < \varepsilon\}/\mathbf{P}\{|W(1)| < \varepsilon\} =$$

$$\mathbf{P}\{\tau < 1, |\tilde{W}(1 - \tau) - x| < \varepsilon\}/\mathbf{P}\{|W(1)| < \varepsilon\} =$$

$$\mathbf{P}\{\tau < 1, |W(1) - 2x| < \varepsilon\}/\mathbf{P}\{|W(1)| < \varepsilon\} =$$

$$\mathbf{P}\{|W(1) - 2x| < \varepsilon\}/\mathbf{P}\{|W(1)| < \varepsilon\} = \frac{\Phi(2x + \varepsilon) - \Phi(2x - \varepsilon)}{\Phi(\varepsilon) - \Phi(-\varepsilon)}$$

So

$$\lim_{\varepsilon \downarrow 0} P(x, \varepsilon) = e^{-2x^2}$$

Since $B$ and $W(1)$ are independent, we have

$$P(x) = \mathbf{P}\{\exists t \in [0,1] : B(t) = x | |W(1)| < \varepsilon\} \le$$

$$\mathbf{P}\{\exists t \in [0,1] : B(t) \ge x - \varepsilon - tW(1) | |W(1)| < \varepsilon\} = P(x - \varepsilon, \varepsilon)$$

Analogously, $P(x) \ge P(x + \varepsilon, \varepsilon)$.

Fix $\delta > \varepsilon > 0$. If $\varepsilon \downarrow 0$ then

$$P(x) \le P(x - \varepsilon, \varepsilon) \le P(x - \delta, \varepsilon) \to e^{-2(x-\delta)^2}$$

$$P(x) \ge P(x + \varepsilon, \varepsilon) \ge P(x + \delta, \varepsilon) \to e^{-2(x+\delta)^2}$$

so

$$e^{-2(x+\delta)^2} \le P(x) \le e^{-2(x-\delta)^2}$$

Going to the limit as $\delta \downarrow 0$ we get $P(x) = e^{-2x^2}$.
$\triangle$

**Property 4.** *For all $x > 0$*

$$\mathbf{P}\{\sup_{0 \le t \le 1} |B_t| \ge x\} = 2 \sum_{n=1}^{\infty} (-1)^{n-1} e^{-2n^2 x^2}. \qquad \text{[B.10]}$$

**Proof.** Let us consider the event

$$A_n(x) = \{\exists 0 \le t_1 < \cdots < t_n \le 1 : B(t_j) = (-1)^{j-1} x, j = 1, \ldots, n\}$$

and the random variables $\tau = \inf\{t : B(t) = x\}$ $\tau' = \inf\{t : B(t) = -x\}$. Set

$$P_n(x) = \mathbf{P}\{A_n(x)\}, \quad Q_n(x) = \mathbf{P}\{A_n(x), \tau < \tau'\}$$

We obtain

$$Q_n(x) + Q_{n+1}(x) = \mathbf{P}\{A_n(x), \tau < \tau'\} + \mathbf{P}\{A_{n+1}(x), \tau < \tau'\} =$$

$$= \mathbf{P}\{A_n(x), \tau < \tau'\} + \mathbf{P}\{A_n(x), \tau' < \tau\} = P_n(x)$$

Property 3 implies that $P_1(x) = e^{-2x^2}$.

Let us find $P_2(x)$. By the formula of complete probability

$$\mathbf{P}\{\exists 0 < t_1 < t_2 \le 1 : W(t_1) = x, W(t_2) = -x, |W(1)| < \varepsilon\} =$$

$$\mathbf{P}\{\exists 0 \le t_1 < t_2 \le 1 : B(t_1) = x - t_1 W_1, B(t_2) =$$

$$-x - t_2 W_1, |W(1)| < \varepsilon\} =$$

$$\int_{-\varepsilon}^{\varepsilon} \mathbf{P}\{\exists 0 < t_1 < t_2 \le 1 : B(t_1) = x - t_1 v, B(t_2) =$$

$$-x - t_2 v\}\varphi(v)dv$$

where $\varphi(v)$ is the probability density of the standard normal distribution. So

$$\lim_{\varepsilon \downarrow 0} \frac{1}{2\varepsilon}\mathbf{P}\{\exists 0 < t_1 < t_2 \le 1 : W(t_1) = x, W(t_2) = -x \mid |W(1)| < \varepsilon\} =$$

$$P_2(x) \lim_{\varepsilon \downarrow 0} \frac{\Phi(\varepsilon) - \Phi(-\varepsilon)}{2\varepsilon} = P_2(x)\varphi(0)$$

By the reflection rule, the probability in the left side can be written as follows

$$\mathbf{P}\{\exists 0 < t_1 < t_2 \le 1 : W(t_1) = x, W(t_2) = -x, |W(1)| < \varepsilon\} =$$

$$\mathbf{P}\{\exists 0 < t_1 < t_2 < 1 : W(t_1) = x, W(t_2) = 3x, |W(1) - 4x| < \varepsilon\} =$$

$$\mathbf{P}\{|W(1) - 4x| < \varepsilon\} = \Phi(4x + \varepsilon) - \Phi(4x - \varepsilon)$$

So we obtain

$$P_2(x) = \lim_{\varepsilon \downarrow 0} \frac{\Phi(4x + \varepsilon) - \Phi(4x - \varepsilon)}{2\varepsilon\varphi(0)} = e^{-8x^2}$$

Analogously, $P_n(x) = e^{-2n^2x^2}$.

We obtain

$$Q_1 = P_1 - Q_2 = P_1 - P_2 + Q_3 = \sum_{k=1}^{n-1} (-1)^k P_k + (-1)^n Q_n$$

Since $Q_n \le P_n \to 0$, we have

$$Q_1(x) = \sum_{n=1}^{\infty} (-1)^n e^{-2n^2x^2}$$

Therefore

$$\mathbf{P}\{ \sup_{0 \le t \le 1} |B_t| \ge x \} = 2Q_1(x) = 2\sum_{n=1}^{\infty} (-1)^n e^{-2n^2x^2}$$

$\triangle$

# Bibliography

[AGU 94a] AGUIREE N., NIKULIN, M., "Chi squared goodness-of-fit test for the family of logistic distributions", *Kybernetika*, vol. 30, 3, p. 214–222,1994.

[AGU 94b] AGUIREE, N., NIKULIN, M., "Goodness-of-fit test for the family of logistic distributions", *Questiio*, vol. 18, 3, p. 317–335, 1994.

[AND 52] ANDERSON, T.W., DARLING, D.A.,"Asymptotic theory of certain goodness-of-fit criteria based on stochastic processes", *Ann. of Statist.*, vol. 23, p. 193–212, 1952.

[AND 54] ANDERSON, T.W., DARLING, D.A., "A test on goodness of fit", *J. Amer. Statist. Assoc.*, vol. 49, p. 765–769, 1954.

[AND 62] ANDERSON, T.W., "On the distribution of the two-sample Cramér–von-Mises criterion", *Ann. Math. Statist.*, vol. 33, p. 1148–1159, 1962.

[ANS 60] ANSARI, A.R., BRADLEY, R.A., "Rank-sum tests for dispersion", *Ann. Math. Statist.*, vol. 31, p. 1174–1189, 1960.

[ASC 90] ASCHER, S., "A survey of tests for exponentiallity", *Comm. Statist. – Theory and Methods*, vol. 19, p. 1811–1825, 1990.

[BAG 10] BAGDONAVIČIUS, V., KRUPIOS, J., NIKULIN, M., *Nonparametric Tests for Censored Data*, ISTE: London, 2011.

[BAK 09] BAKSHAEV, A., "Goodness of fit and homogeneity tests on the basis of N-distances", *J. Statist. Plann. Infer.*, vol. 139, p. 3750–3758, 2009.

[BAK 10] BAKSHAEV, A., "N-distance tests for composite hypotheses of goodness of fit", *Lithuanian Math. Journal*, vol. 50, p. 14–34, 2010.

[BAL 95] BALAKRISHNAN, N., BASU, A.P., *The Exponential Distribution: Theory, Methods and Applications*, OPA: Amsterdam, 1995.

[BAR 82] BARTELS, R., "The rank version of Von Neumann's ratio test for randomness", *J. Amer. Statist. Assoc.*, vol. 77, p. 40–46, 1982.

[BAR 88] BARLOW, W., PRENTICE, R., "Residuals for relative risk regression", *Biometrika*, vol. 75, p. 65–74, 1988.

[BAR 89] BARINGHAUS, L., DANSCHKEA, R., HENZEA, N., "Recent and classical tests for normality – a comparative study", *Comm. Statist.-Simulation*, vol. 18. p. 363–379, 1989.

[BAR 92] BARINGHAUS, N., HENZE, L., "A goodness-of-fit test for the Poisson distribution based on the empirical generating function", *Statist. Probab. Lett.*, vol. 13., p. 269–274, 1992.

[BES 85] BEST, D.J., RAYNER, J.C., "Lancaster's test of normality", *J. Statist. Plan. Inf.*, vol. 12, p. 395–400, 1985.

[BIL 79] BILLINGSLEY, P., *Probability and Measure*, Wiley: New York, 1979.

[BOL 65] BOLSHEV, L.N., "On characterization of the Poisson distribution and its statistical applications", *Theory of Probability and its Applications*, vol. 10, p. 488–499, 1965

[BOL 66] BOLSHEV, L.N., "On goodness-of-fit for exponential distribution", *Theory of probability and its applications*, vol. 11, p. 542–544, 1966.

[BOL 78] BOLSHEV, L.N., MIRVALIEV, M., "Chi-square goodness-of-fit tests for the Poisson, binomial, and negative binomial distributions", *Theory of Probability and its Applications*, vol.23, p. 461–474, 1978.

[BOL 83] BOLSHEV, L.N., SMIRNOV, N.N., *Tables of Mathematical Statistics*, Nauka: Moscow, 1983.

[BOL 87] BOLSHEV, L.N., *Selected papers. Theory of Probability and Mathematical Statistics*, Nauka: Moscow, p. 286, 1987.

[BOW 75]  BOWMAN, K., SHENTON, L.R., "Omnibus contours for departures from normality based on b1 and b2", *Biometrika*, vol. 63, p. 243–250, 1975.

[BRO 60]  BROWNLEE, K.A., *Statistical Theory and Methodology in Science and Engineering*, Wiley: New York, 1960

[CAB 05]  CABANA, A., QUIROZ, A.J., "Using the empirical moment generating function in testing the Weibull and the Type I extreme value distributions", *Test*, vol. 14., p. 417–431, 2005.

[CHE 04]  ČEKANAVIČIUS, V., MURAUSKAS, G., *Statistics and its Applications II*, TEV, Vilnius, 2004.

[COC 54]  COCHRAN, W.G., "Some methods for strengthening the common chi square tests", *Biometrics*, vol. 10, pp 417–451, 1954.

[CON 71]  CONNOVER, W.J., *Practical Non-parametric Statistics*, Wiley: New York, 1971.

[COR 09]  CORDER, G.W., FOREMAN, D.I., *Nonparametric Statistics for Non-Statisticians: A Step-by-Step Approach*, Wiley: New Jersey, 2009.

[CRA 46]  CRAMÉR, H., *Mathematical Methods of Statistics*, Princeton University Press, 1946.

[DAG 70]  D'AGOSTINO, R.B., "Transformation to normality of the null distribution of g1", *Biometrika*, vol. 57, pp. 679–681, 1970.

[DAG 71]  D'AGOSTINO, R., "An omnibus test of normality for moderate and large samples", *Biometrika*, vol. 58, p. 341–348, 1971.

[DAG 86]  D'AGOSTINO, R.B., STEPHENS, M.A., *Goodness-of-fit techniques*, Marcel Dekker: New York, 1986.

[DAG 90]  D'AGOSTINO, R., BELANGER, A., "An omnibus test of normality for moderate and large samples", *American Statistician*, vol. 44, p. 316–322, 1990.

[DAR 55]  DARLING, D.A., "Cramer–Smirnov test in the parametric case", *Ann. Math. Statist.*, vol. 26, 1955.

[DAR 57]  DARLING, D.A., "The Kolmogorov–Smirnov, Cramér–von-Mises tests", *Ann. Math. Statist.*, vol. 28, p. 823–838, 1957.

[DAR 83a] DARLING, D., "On the supremum of a certain Gaussian process", *Ann. Statist.*, vol. 11, p. 803–806, 1983.

[DAR 83b] DARLING, D., "On the asymptotic distribution of Watson's statistics", *Ann. Statist.*, vol. 11, p. 1263–1266, 1983.

[DRO 88] DROST, F., *Asymptotics for Generalized Chi-square Goodness-of-fit Tests*, Amsterdam: Center for Mathematics and Computer Sciences, CWI Tracts, vol. 48, 1988.

[DRO 89] DROST, F., "Generalized chi square goodness-of-fit tests for location-scale models when the number of classes tend to infinity", *Ann. Statist.*, vol. 17, 1989.

[DUD 79] DUDLEY, R.M., "On $\chi^2$ tests of composite hypotheses", in: *Probability Theory*, Banach Center Publications: Warsaw, vol.5, p. 75–87, 1979.

[DUR 73] DURBIN, J., "Weak convergence of the sample distribution function when parameters are estimated", *Ann. Math. Statist.*, vol. 1, p. 279–290, 1973.

[DZH 74] DZHAPARIDZE, K.O., NIKULIN, M.S., "On a modification of the standard statistics of Pearson", *Theory of Probability and its Applications*, vol. 19, 4, p. 851–852, 1974.

[EPP 83] EPPS, T.W., PULLEY, L.B., "A test for normality based on the empirical characteristic function", *Biometrika*, vol. 70, p. 723–726, 1983.

[FIS 50] FISHER, R.A., *Contributions to Mathematical Statistics*, Wiley: New York, 1950.

[FLE 81] FLEMING, T., HARRINGTON. D., "A class of hypothesis tests for one and two samples of censored survival data", *Comm. Statist.*, vol. 10, p. 763–794, 1981.

[FRI 37] FRIEDMAN, M., "The use of ranks to avoid the assumption of normality implicit in the analysis of variance", *J. Amer. Statist. Assoc.*, vol. 32, p. 675–701, 1937.

[FRO 78] FROSINI, B.V., "A survey of a class of goodness-of-fit statistics", *Metron.*, vol. 36, p. 3–49, 1978.

[FRO 87] FROSINI, B.V., "On the distribution and power of goodness-of-fit statistic with parametric and nonparametric applications", *Goodness-of-fit*, eds. Revesz, P., Sarkadi, K., Sen, P.K., North Holland: Amsterdam, 1987, p. 133–154, 1987.

[GEA 35] GEARY, R.C., "The ratio of the mean deviation to the standard deviation as a test of normality", *Biometrika*, vol. 27, p. 310–322, 1935.

[GIB 09] GIBBONS, J.D., CHAKRABORTI, S., *Nonparametric Statistical Inference*, CRC Press, 5th edn., 2009.

[GNE 65] GNEDENKO, B.V., BELYAEV, Y.K., SOLOVIEV, A.D., *Mathematical Methods in Reliability*, Nauka: Moscow, 1965.

[GOV 07] GOVINDARAJULU, Z., *Nonparametric Inference*, World Scientific, 2007.

[GRE 96] GREENWOOD, P.E. AND NIKULIN, M., *A Guide to Chi-squared Testing*, Wiley: New York, 1996.

[HAJ 67] HÁJEK, J., SIDÁK, Z., *Theory of Rank Tests*, Academic Press: New York, 1967.

[HAJ 99] HÁJEK, J., SIDÁK, Z., SEN, P.K., *Theory of Rank Tests*, Academic Press: San Diego, 2nd edn., 1999.

[HEG 75] HEGAZY, Y.A., GREEN, J.R., "Some new goodness-of-fit tests using order statistics", *Applied Statistics*, vol. 24, p. 299–308, 1975.

[HET 84] HETTMANSPERGER, T.P., *Statistical Inference based on Ranks*, Wiley: New York, 1984.

[HOL 99] HOLLANDER, M., WOLFE, D.A., *Nonparametric Statistical Methods*, 2nd edn., Wiley: New York, 1999.

[JAR 87] JARQUE, C.M., BERA, A.K., "A test for normality of observations and regression residuals", *Int. Stat. Rev.*, vol. 55, p. 163–172, 1987.

[KAC 55] KAC, M., KIEFER, J., WOLFOWITZ, J., "On tests for normality and other tests of goodness of fit based on distance methods", *Ann. Math. Statist.*, vol. 26, p. 189–211, 1955.

[KAL 89] KALBFLEISCH, J., PRENTICE, R., *The Statistical Analysis of Failure Time Data*, Wiley: New York, 1980.

[KEN 38] KENDALL, M., "A New Measure of Rank Correlation", *Biometrika*, vol.30, p. 81–89, 1938.

[KEN 39] KENDALL, M.G., BABINGTON SMITH, B., "The problem of m rankings", *Ann. Math. Statist.*, vol. 10, p. 275–287, 1939.

[KEN 62] KENDALL, M.G., "Rank Correlation Methods", Griffin, 1962.

[KHM 77] KHMALADZE, E., "On omega-square tests for parametric hypotheses", *Theory of Probability and its Applications*, vol. 26, p. 246–265, 1977.

[KLO 65] KLOTZ, J., "Alternative efficiencies for signed rank tests", *Ann. Math. Statist.*, vol. 36, 6, p. 1759–1766, 1965.

[KOL 33] KOLMOGOROV, A.N., "Sulla determinizione empirica di una legge di distribuzione", *Giorn. Ist. Ital. Attuari*, vol. 4, p. 83–91, 1933.

[KOL 67] KOLCHIN, V.F., SEVASTIANOV, B.A., CHISTIAKOV, V.N., *Random*, Nauka: Moscow, 1967.

[KRU 52] KRUSKAL, W.H., WALLIS, W.A., "Use of ranks in one-criterion variance analysis", *J. Amer. Statist. Assoc.*, vol. 47, p. 583–621, 1952.

[KUI 60] KUIPER, N., "Tests concerning random points on the circle", *Proc. Kon. Ned. Akad. van Wet*, vol. 60, p. 38–47, 1960.

[LAW 02] LAWLESS, J.F., *Statistical Models and Methods for Lifetime Data*, Wiley, 2nd edn., 2002.

[LEC 83] LECAM, L., MAHAN, C., SINGH, A., "An extension of a theorem of H. Chernoff and E.L. Lehmann", in *Recent Advances in Statistics*, Academic Press, Orlando, p. 303–332, 1983.

[LEH 51] LEHMANN, E., "Consistency and unbiasedness of certain nonparametric tests", *Ann. Math. Statist.*, vol. 22, p. 165–179, 1951.

[LEM 98a] LEMESHKO, B.YU., POSTOVALOV, S.N., "On the distributions of nonparametric goodness-of-fit test statistics when estimating distribution parameters by samples", *Ind. Lab. Math. Diag.*, vol. 64, 3, p. 61–72, 1998.

[LEM 98b] LEMESHKO, B.YU., "Asymptotically optimal grouping of observations in goodness-of-fit tests", *Ind. Lab. Math. Diag.*, vol. 64, 1, p. 56–64, 1998.

[LEM 98c] LEMESHKO, B.YU., POSTOVALOV, S.N., "On the dependence of the limiting statistic distributions of the Pearson chi-squire and the likelihood ratio tests upon the grouping method", *Ind. Lab. Math. Diag.*, vol. 64, 5, p. 56–63, 1998.

[LEM 01] LEMESHKO, B.YU., POSTOVALOV, S.N., CHIMITOVA, E.V., "On statistic distributions and the power of the Nikulin chi-square test", *Ind. Lab. Math. Diag.*, vol. 67, 3, p. 52–58, 2001.

[LEM 02] LEMESHKO, B.YU., CHIMITOVA, E.V., "Errors and incorrect procedures when utilizing $\chi^2$ fitting criteria", *Measurement Techniques*, vol.45, 6, p. 572–581, 2002.

[LEM 03] LEMESHKO, B.YU., CHIMITOVA, E.V., "On the choice of the number of intervals in chi-square goodness-of-fit tests", *Ind. Lab. Math. Diag.*, vol. 69, 1, p. 61–67, 2003.

[LEM 04] LEMESHKO, B.YU., "Errors when using nonparametric fitting criteria", *Measurement Techniques*, vol. 47, 2, p. 134–142, 2004.

[LEM 05] LEMESHKO, B.YU., LEMESHKO, S.B., "A comparative analysis of tests for deviations of probabilities distribution from normal distribution", *Metrology*, vol. 2, p. 3–24, 2005.

[LEM 09a] LEMESHKO, B.YU., LEMESHKO, S.B., "Distribution models for nonparametric tests for fit in verifying complicated hypotheses and maximum-likelihood estimators. Part 1", *Measurement Techniques*, vol. 52, 6, p. 555–565, 2009.

[LEM 09b] LEMESHKO, B.YU., LEMESHKO, S.B., "Models for statistical distributions in nonparametric fitting tests on composite hypotheses based on maximum-likelihood estimators. Part II", *Measurement Techniques*, vol. 52, 8, p. 799–812, 2009.

[LEM 10] LEMESHKO, B.YU, LEMESHKO, S.B., POSTOVALOV, S.N., "Statistic distribution models for some nonparametric goodness-of-fit tests in testing composite hypotheses", *Communications in Statistics. Theory and Methods*, vol. 39, 3, p. 460-471, 2010.

[LIL 67] LILLIEFORS, H.W., "On the Kolmogorov-Smirnov test for normality with mean and variance unknown", *J. Amer. Statist. Assoc.*, vol. 62, p. 399-402, 1967.

[LIL 69] LILLIEFORS, H.W., "On the Kolmogorov-Smirnov test for normality with mean and variance unknown", *J. Amer. Statist. Assoc.*, vol. 64, p. 387-389, 1969.

[LIN 80] LIN, C.C., MUDHOLKAR, G.S., "A simple test for normality against asymmetric alternatives", *Biometrika*, vol. 67, p. 455-461, 1980.

[LOC 76] LOCKE, C., SPURRIER, J.D., "The use of u-statistics for testing normality against nonsymmetric alternatives", *Biometrika*, vol. 63, p. 143–147, 1976.

[MAA 68] MAAG, U., STEPHENS, M., "The v(n,m) two-sample test", *Ann. Math. Statist.*, vol. 39, p. 923–935, 1968.

[MAN 47] MANN, H.B., WHITNEY, D.R., "On a test of whether one of two random variables is stochastically larger than the other", *Ann. Math. Statist.*, vol. 18, p. 50–60, 1947.

[MAN 73] MANN, N.R., SCHEUER, E.M., FERTIG, K.W., "A new goodness-of-fit test for the two parameter Weibull or extreme-value distribution with unknown parameters", *Communications in Statistics*, vol. 2, p. 383–400, 1973.

[MAR 75] MARTYNOV, G.V., "Computation of the distribution functions of quadratic forms of normal random variables", *Theory of Probability and its Applications*, vol. 20, p. 797–809, 1975.

[MAR 77] MARTYNOV, G.V., *Omega-Square Criteria*, Nauka: Moscow, 1977.

[MAR 95] MARITZ, J.S., *Distribution-Free Statistical Mehods*, Chapman & Hall/CRC: Boca Raton, 1995.

[MAR 97] MARZEC, L., MARZEC, P., "Generalized martingale-residual processes for goodness-of-fit inference in Cox's type regression models", *Ann. Statist.*, vol. 25, p. 683–714, 1997.

[MCN 47] MCNEMAR, Q., "Note on the sampling error of the difference between correlated proportions or percentages", *Psychometrika*, vol. 12, p. 153–157, 1947.

[MIS 31] MISES, R. VON., *Warhrscheinlichkeit, Statistik und Wahrheit*, Springer-Verlag, 1931.

[MOO 54] MOOD, A.M., "On the asymptotic efficiency of certain nonparametric two sample tests", *Ann. Math. Statist.*, vol. 25, p. 514–522, 1954.

[NIK 73a] NIKULIN, M.S.,"Chi-square test for normality", in *Proceedings of the International Vilnius Conference on Probability Theory and Mathematical Statistics*, vol. 2, p. 119–122, 1973.

[NIK 73b] NIKULIN, M.S., "Chi-square test for continuous distributions with shift and scale parameters", *Theory of Probability and its Applications*, vol. 18, p. 559–568, 1973.

[NIK 73c] NIKULIN, M.S., "On a chi-square test for continuous distributions", *Theory of Probability and its Applications*, vol. 18, 3, p. 638–639, 1973.

[NIK 92] NIKULIN, M., MIRVALIEV, M., "Goodness-of-fit chi-squared type criterions", *Industrial Laboratory*, p. 280–291, 1992.

[NIK 95] NIKITIN, Y., *Asymptotic Efficiency of Nonparametric Tests*, Cambridge University Press: New York, 1995.

[PAR 99] PARK, S., "A goodness-of-fit test for normality based on the sample entropy of order statistics", *Statist. Prob. Lett.*, vol. 44, p. 359–363, 1999.

[PEA 48] *Karl Pearson's Early Statistical Papers*, Cambridge University Press: London, 1948.

[PEA 00] PEARSON, K., "On the criterion that a given system of deviation from the probable in the case of a correlated system of variables is such that it can be reasonably supposed to have arisen from random sampling", *Phil. Mag.*, vol.50, p. 157–175, 1900.

[PET 76] PETTITT, A., "A two sample Anderson–Darling rank statistics", *Biometrika*, vol. 63, p. 161–168, 1976.

[PET 79] PETTITT, A., "A two sample Cramer–von-Mises type rank statistics", *J. Roy. Statist. Soc., A*, vol. 142, p. 46–53, 1979.

[PRE 76] PRESCOTT, P., "On test for normality based on sample entropy", *J. Roy. Statist. Soc.*, vol. 38, p. 254–256, 1976.

[PIT 48] PITMAN, E.J.G., Non-parametric Statistical Inference, University of North Carolina Institute of Statistics (lecture notes), 1948.

[RAO 74] RAO K.C., ROBSON D.S., "A chi-squared statistic for goodness-of-fit tests within the exponential distribution", *Commun. Statist.*, vol. 3, p. 1139–1153, 1974.

[RAO 02] RAO, C.R., *Linear Statistical Inference and its Applications*, Wiley: New York, 2nd edn., 2002.

[RIN 09] RINNE, H., *The Weibull Distribution: A Handbook*, CRC Press, 2009.

[ROS 52] ROSENBLATT, M., "Limit theorems associated with variants of the Von Mises statistics", *Ann. Math. Statist.*, vol. 23, p. 617–623, 1952.

[RUT 30] RUTHERFORD, E., CHADWICK, J., ELLIS, C.D., *Radiation from Radioactive Substances*, Cambridge University Press: London, 1930.

[SAR 81] SARKADI, K., "The asymptotic distribution of certain goodness of fit test statistics", *The First Pannon Symposium on Math. Statist.*, ed. W. Wertz, Lecture Notes in Statistics. vol. 8, Springer-Verlag: New York, p. 245–253, 1981.

[SHE99] SCHEFFÉ, H., *The Analysis of Variance*, Wiley: New York, 1999.

[SHA 65] SHAPIRO, S.S., WILK, M.B., "An analysis of variance test for normality", *Biometrika*, vol. 52, p. 591–612, 1965.

[SHA 72] SHAPIRO, S.S., FRANCIA, R.S., "An approximate analysis of variance test for normality", *J. Amer. Statist. Assoc.*, vol. 67, p. 215–216, 1972.

[SIE 60] SIEGEL S., TUKEY, J.W., "A nonparametric sum of ranks procedure for relative spread in unpaired samples", *J. Amer. Statist. Assoc.*, vol. 55, p. 429–445, 1965.

[SMI 36] SMIRNOV, N.V., "Sur la distribution de $\omega^2$", *C. R. Acad. Sci. Paris*, vol. 202, p. 449–452, 1936.

[SMI 37] SMIRNOV, N.V., "On the distribution of Mises $\omega^2$-test", *Math. Sbornik*, vol. 2, p. 973–994, 1937.

[SMI 39a] SMIRNOV, N.V., "On estimating the discrepancy between empirical distribution functions in two independent samples", *Bulletin of Moscow Gos. University*, Ser. A, vol. 2, p. 3–14, 1939.

[SMI 39b] SMIRNOV, N.V., "On deviation of the empirical distribution function", *Math. Sbornik*, vol. 6, p. 3–26, 1939.

[SMI 44] SMIRNOV, N.V., "Approximate distribution laws for random variables, constructed from empirical data", *Uspekhi Math. Nauk*, vol. 10, p. 179–206, 1944.

[SPE 04] SPEARMAN, C., "The proof and measurement of association between two things", *Amer. J. Psychol.*, vol. 15, p. 72–101, 1904.

[SPI 77] SPIEGELHALTER, O.J., "A test for normality against symmetric alternatives", *Biometrika*, vol. 64, p. 415–418, 1977.

[SPR 01] SPRENT, P., SMEETON, N.C., *Applied Nonparametric Statistical Methods*, Chapman & Hall/CRC: Boca Raton, 2001.

[STE 74] STEPHENS, M.A., "EDF statistics for goodness of fit and some comparisons", *J. Amer. Statist. Assoc.*, vol. 69, p. 730–737, 1974.

[STE 76] STEPHENS, M.A., "Asymptotic results for goodness-of-fit statistics with unknown parameters", *Ann. Statist.*, vol. 4, p. 357–369, 1976.

[STE 77] STEPHENS, M.A., "Goodness of fit for the extreme value distribution", *Biometrika*, vol. 64, p. 583–588, 1977.

[STE 79] STEPHENS, M.A., "Tests of fit for the logistic distribution based on the empirical distribution function", *Biometrika*, vol. 66, p. 591–595, 1979.

[STU 93] STUTE, W., GONZALES–MANTEIGA, W., PRESEDP–QUINDIMIL, M., "Bootstrap based goodness-of-fit tests", *Metrika*, vol. 40, p. 243–256, 1993.

[SZU 08] SZUCS, G., "Parametric bootstrap tests for continuous and and discrete distributions", *Metrika*, vol. 67, p. 33–81, 2008.

[TYU 70] TYURIN, Y.N., "On testing parametric hypotheses by nonparametric methods", *Theory of Probability and its Applications*, vol. 25, p. 745–749, 1970.

[TYU 84] TYURIN, Y.N., "On the limit distribution of Kolmogorov–Smirnov statistics for a composite hypothesis", *Izv. Akad. Nauk SSSR*, vol. 48, p. 1314–1343, 1984.

[VAN 52] VAN DER VAERDEN, B.L., "Order tests for the two-sample problem and their power", *Proc. Kon. Ned. Akad. Wetensch., A*, vol. 55, p. 453–458, 1952.

[VAN 00] VAN DER VAART, A.W., *Asymptotic Statistics*, Cambridge University Press, 2000.

[WAL 40] WALD, A., WOLFOWITZ, J., "On a test whether two samples are from the same population", *Ann. Math. Statist.*, vol. 11, p. 147–162, 1940.

[WAT 40] WATSON, G., "On the test whether two samples are from the same population", *Ann. Math. Statist.*, vol. 11, p. 147–162, 1940.

[WAT 76] WATSON, G., "Optimal invariant tests for uniformity", *Studies in Probability and Statistics. Paper in honour of E.J. Pitman*, p. 121–127, 1976.

[WIL 45] WILCOXON, F., "Individual comparisons by ranking methods", *Biometrics*, vol. 1, p. 80–83, 1945.

[WIL 47] WILCOXON, F., "Probability tables for individual comparisons by ranking methods", *Biometrics*, vol. 3, p. 119, 1947.

[YAT 34] YATES, F., "Contingency table involving small numbers and the chi square test", *Supplement to the Journal of the Royal Statistical Society*, vol. 1, p. 217–235, 1934.

[ZHA 99b] ZHANG B., "Omnibus test for normality using the q-statistic", *Journal of Appl.Statist.*, vol. 26, p. 519–528, 1999.

# Index